T0183334

Communications
in Computer and Information Science 1285

Commenced Publication in 2007
Founding and Former Series Editors:
Simone Diniz Junqueira Barbosa, Phoebe Chen, Alfredo Cuzzocrea,
Xiaoyong Du, Orhun Kara, Ting Liu, Krishna M. Sivalingam,
Dominik Ślęzak, Takashi Washio, Xiaokang Yang, and Junsong Yuan

More information about this series at http://www.springer.com/series/7899

Gabriele Kotsis · A Min Tjoa ·
Ismail Khalil · Lukas Fischer ·
Bernhard Moser · Atif Mashkoor ·
Johannes Sametinger · Anna Fensel ·
Jorge Martinez-Gil (Eds.)

Database and Expert Systems Applications

DEXA 2020 International Workshops
BIOKDD, IWCFS and MLKgraphs
Bratislava, Slovakia, September 14–17, 2020
Proceedings

 Springer

Editors
Gabriele Kotsis
Johannes Kepler University of Linz
Linz, Austria

A Min Tjoa
Vienna University of Technology
Vienna, Wien, Austria

Ismail Khalil
Johannes Kepler University of Linz
Linz, Oberösterreich, Austria

Lukas Fischer
Software Competence Center Hagenberg
Linz, Austria

Bernhard Moser
Software Competence Center Hagenberg
Linz, Austria

Atif Mashkoor
Software Competence Center Hagenberg
Linz, Austria

Johannes Sametinger
Johannes Kepler University of Linz
Linz, Austria

Anna Fensel
University of Innsbruck
Innsbruck, Tirol, Austria

Jorge Martinez-Gil
Software Competence Center Hagenberg
Linz, Austria

ISSN 1865-0929   ISSN 1865-0937   (electronic)
Communications in Computer and Information Science
ISBN 978-3-030-59027-7   ISBN 978-3-030-59028-4   (eBook)
https://doi.org/10.1007/978-3-030-59028-4

This Springer imprint is published by the registered company Springer Nature Switzerland AG
The registered company address is: Gewerbestrasse 11, 6330 Cham, Switzerland

Preface

The Database and Expert Systems Applications (DEXA) workshops are a platform for the exchange of ideas, experiences, and opinions among scientists and practitioners – those who are defining the requirements for future systems in the areas of database and artificial technologies.

This year DEXA featured three international workshops:

- BIOKDD 2020 – The 11th International Workshop on Biological Knowledge Discovery from Data
- IWCFS 2020 – The 4th International Workshop on Cyber-Security and Functional Safety in Cyber-Physical Systems
- MLKgraphs 2020 – The Second International Workshop on Machine Learning and Knowledge Graphs

The DEXA workshops papers included papers that focus mainly on very specialized topics such as applications of database and expert systems technology.

We would like to thank all workshop chairs and Program Committee members for their excellent work, namely Lukas Fischer and Bernhard Moser, the co-chairs of the BIOKDD workshop; Atif Mashkoor and Johannes Sametinger, the co-chairs of the IWCFS workshop; and Anna Fensel, Bernhard Moser, and Jorge Martinez-Gil, the co-chairs of the MLKgraphs workshop.

DEXA 2020 was the 31st in the series of annual scientific platform on database and expert systems applications after Vienna, Berlin, Valencia, Prague, Athens, London, Zurich, Toulouse, Vienna, Florence, Greenwich, Munich, Aix en Provence, Prague, Zaragoza, Copenhagen, Krakow, Regensburg, Turin, Linz, Bilbao, Toulouse, Vienna, Prague, Munich, Valencia, Porto, Lyon, Regensburg, and Linz.

This year DEXA was very unique. Due to the pandemic and for the safety of all participants as well as other restrictions preventing travel and gatherings, this year DEXA was held as a virtual conference in the Western European time zone.

We would like to express our thanks to all institutions actively supporting this event, namely:

- Johannes Kepler University Linz (JKU)
- Software Competence Center Hagenberg (SCCH)
- The International Organization for Information Integration and Web based applications and Services (@WAS)

Finally, we hope that all the participants' of DEXA 2020 workshops enjoyed the program that we put together.

September 2020

Gabriele Kotsis
A Min Tjoa
Ismail Khalil

Organization

Steering Committee

Gabriele Kotsis	Johannes Kepler University Linz, Austria
A Min Tjoa	Technical University of Vienna, Austria
Ismail Khalil	Johannes Kepler University Linz, Austria

BIOKDD 2020 Chairs

Lukas Fischer	Software Competence Center Hagenberg, Austria
Bernhard Moser	Software Competence Center Hagenberg, Austria

BIOKDD 2020 Program Committee and Reviewers

Jamal Al Qundus	FU Berlin, Germany
Matteo Comin	University of Padova, Italy
Suresh Dara	BV Raju Institute of Technology, India
Tomas Flouri	Heidelberg Institute for Theoretical Studies, Germany
Manuela Geiss	Software Competence Center Hagenberg, Austria
Adrien Goeffon	LERIA, Université d'Angers, France
Robert Harrison	Georgia State University, USA
Daisuke Kihara	Purdue University, USA
Mohit Kumar	Software Competence Center Hagenberg, Austria
Dominique Lavenier	CNRS, IRISA, France
Vladimir Makarenkov	Université du Québec à Montréal, Canada
Mirto Musci	University of Pavia, Italy
Solon Pissis	Centrum Wiskunde & Informatica (CWI), The Netherlands
Aruna Rao	BVRITH College of Engineering for Women, India
Maad Shatnawi	United Arab Emirates University, UAE
Stefan Thumfart	RISC Software GmbH, Austria
Emanuel Weitschek	Uninettuno International University, Italy
Malik Yousef	Zefat Academic College, Israel

IWCFS 2020 Chairs

Atif Mashkoor	Software Competence Center Hagenberg, Austria
Johannes Sametinger	Johannes Kepler University Linz, Austria

IWCFS 2020 Program Committee and Reviewers

Yamine Ait Ameur	IRIT, INPT-ENSEEIHT, France
Paolo Arcaini	National Institute of Informatics, Japan

Richard Banach	The University of Manchester, UK
Ladjel Bellatreche	LIAS, ENSMA, France
Jorge Cuellar	Siemens AG, Germany
Alexander Egyed	Johannes Kepler University Linz, Austria
Osman Hasan	National University of Sciences and Technology, Canada
Irum Inayat	National University of Computers and Emerging Sciences, Pakistan
Jean-Pierre Jacquot	LORIA, Henri Poincaré University, France
Muhammad Taimoor Khan	University of Greenwich, UK
Martín Ochoa	AppGate Inc., Colombia
Tope Omitola	University of Southampton, UK
Neeraj Singh	University of Toulouse, France

MLKgraphs 2020 Chairs

Anna Fensel	University of Innsbruck, Austria
Bernhard Moser	Software Competence Center Hagenberg, Austria
Jorge Martinez-Gil	Software Competence Center Hagenberg, Austria

MLKgraphs 2019 Program Committee and Reviewers

Anastasia Dimou	Ghent University, Belgium
Lisa Ehrlinger	Johannes Kepler University Linz and Software Competence Center Hagenberg, Austria
Agata Filipowska	Poznań University of Economics and Business, Poland
Isaac Lera	University of the Balearic Islands, Spain
Vit Novacek	National University of Ireland Galway, Ireland
Femke Ongenae	Ghent University, Belgium
Mario Pichler	Software Competence Center Hagenberg, Austria
Artem Revenko	Semantic Web Company GmbH, Austria
Marta Sabou	Vienna University of Technology, Austria
Harald Sack	Leibniz Institute for Information Infrastructure and KIT Karlsruhe, Germany
Iztok Savnik	University of Primorska, Slovenia
Marina Tropmann-Frick	Hamburg University of Applied Sciences, Germany
Adrian Ulges	RheinMain University of Applied Sciences, Germany

Organizers

Contents

Biological Knowledge Discovery
from Data

An In-Memory Cognitive-Based Hyperdimensional Approach to Accurately Classify DNA-Methylation Data of Cancer

Fabio Cumbo[1]([⊠]) and Emanuel Weitschek[2]

[1] Department of Cellular, Computational, and Integrative Biology (CIBIO),
University of Trento, Via Sommarive 9, 38123 Povo, Trento, Italy
`fabio.cumbo@unitn.it`
[2] Department of Engineering, Uninettuno University,
Corso Vittorio Emanuele II 39, 00186 Rome, Italy
`emanuel.weitschek@uninettunouniversity.net`

Abstract. With Next Generation DNA Sequencing techniques (NGS) we are witnessing a high growth of genomic data. In this work, we focus on the NGS DNA methylation experiment, whose aim is to shed light on the biological process that controls the functioning of the genome and whose modifications are deeply investigated in cancer studies for biomarker discovery. Because of the abundance of DNA methylation public data and of its high dimension in terms of features, new and efficient classification techniques are highly demanded. Therefore, we propose an energy efficient in-memory cognitive-based hyperdimensional approach for classification of DNA methylation data of cancer. This approach is based on the brain-inspired Hyperdimensional (HD) computing by adopting hypervectors and not single numerical values. This makes it capable of recognizing complex patterns with a great robustness against mistakes even with noisy data, as well as the human brain can do. We perform our experimentation on three cancer datasets (breast, kidney, and thyroid carcinomas) extracted from the Genomic Data Commons portal, the main repository of tumoral genomic and clinical data, obtaining very promising results in terms of accuracy (i.e., breast 97.7%, kidney 98.43%, thyroid 100%, respectively) and low computational time. For proving the validity of our approach, we compare it to another state-of-the-art classification algorithm for DNA methylation data. Finally, processed data and software are freely released at https://github.com/fabio-cumbo/HD-Classifier for aiding field experts in the detection and diagnosis of cancer.

Keywords: Hyperdimensional computing · Energy efficient · DNA-methylation · Cancer · TCGA · Genomic Data Commons

G. Kotsis et al. (Eds.): DEXA 2020 Workshops, CCIS 1285, pp. 3–10, 2020.
https://doi.org/10.1007/978-3-030-59028-4_1

1 Background

Because of its high spread, cancer is one of the most studied diseases in the last decades. It is widely agreed that cellular alteration that leads to the development of tumors is activated by several factors and agents, e.g., physical and biological mutagens, chemicals, bacteria, and viruses. In this regard, the biological process of DNA methylation plays a crucial role, i.e., its modification may regulate the functioning of the genome and interfere in cellular division. DNA Methylation consists in a genetic modification that occurs in human cells and its changes are often related to the development of a disease such as cancer [10]. This means that analyzing DNA methylation data of subjects affected by tumor is a challenge for current cancer research [13]. Thanks to Next Generation Sequencing (NGS) techniques DNA methylation data are widely available at public sources and exponentially growing [11]. Indeed, one of the main repositories of DNA methylation data of cancer is the Genomic Data Commons (http://gdc.cancer.gov/) [6]. It consists of a large collection of genomic and clinical data of more than 83,000 cases and 64 projects. In particular, The Cancer Genome Atlas (TCGA) project stands out, whose aim is to share data and knowledge about cancer since 2005 [12]. Indeed, it contains genomic and clinical data of more than 30 tumor types of over 20,000 patients derived from different NGS experiments, e.g., DNA sequencing, RNA sequencing, DNA methylation. In this work, we focus on DNA methylation genomic data of three cancer types, i.e., Breast Invasive Carcinoma (BRCA), Kidney renal papillary cell carcinoma (KIRP), and Thyroid carcinoma (THCA). Through the application of an in-memory cognitive-based hyperdimensional approach, we investigate the possibility to distinguish tumoral from non-tumoral samples present in the data sets. Indeed, one major challenge for field experts is the early detection and diagnosis of the disease with the aid of clinical and genomic data. Several previous studies dealt with the classification of cancer patients with machine learning techniques [1,2,8,14], but to our knowledge none of them applied brain-inspired Hyperdimensional (HD) computing [7,9] that simulates the human brain by adopting hypervectors and not single numerical values. The term "hyper" refers to the dimensionality of the (pseudo)random vectors, which is typically in the order of the thousands (usually $D = 10{,}000$).

2 Methods

In this work, we take into consideration DNA methylation data of TCGA produced with the Illumina Infinium Human DNA Methylation 450 sequencing platform, which allows to quantify the amount of methylated molecules on more than 450 thousand known CpG regions of the DNA. In TCGA each sample is represented with a list of following fields: gene symbols, chromosomes, genomic coordinates (where the methylations occur), and their methylation values (beta value). The beta value is a measure that shows the percentages of methylated cytosines in a CpG island, i.e., particular regions of the DNA sequence where the methylation occurs. The beta value is the amount of methylated molecules quantified by the sequencing machine on a specific region of the DNA and then normalised according to the Formula 1.

$$\beta_n = \frac{max(Meth_n, 0)}{max(Meth_n, 0) + max(Unmeth_n, 0) + \epsilon} \qquad (1)$$

It is worth noting that the beta value is a continuous variable ranging from 0 up to 1 which means no methylation at all and fully methylated region respectively. In Formula 1, $Meth_n$ and $Unmeth_n$ are the intensities of the nth methylated and unmethylated allele respectively, while ϵ is required to modulate the beta value in case both the methylated and the unmethylated intensities are very low.

We organize DNA methylation data as a matrix, where the rows represents the samples and the columns represent the features, i.e., we consider the beta values of the 450 thousand CpG sites of the DNA sequence. The last column of the matrix encodes the class label of each sample. Indeed the features are the methylated sites and their values represent the percentages of methylated cytosines in a CpG island (beta value - bv). For each dataset, we consider n samples each one with its m features (with m equal to 450,000) and their class labels (conditions), e.g., tumoral and normal. Each element of the matrix contains the beta value associated to the CpG site. In Table 1 we show the structure of the dataset. In order to classify the tumoral and normal samples of DNA methylation data of BRCA, KIRP, and THCA, we redesign a method that was previously adopted for language and speech recognition problems [4,5]. Our goal is to build two HD vectors representing the tumoral and normal classes and implementing the classification model that was derived during the training phase. Thus, for predicting the class of every sample the inner product between their HD representations and the HD vectors of the classes is computed. We summarized the proposed method with the following four steps depicted in Fig. 1.

Table 1. Data matrix of DNA methylation

Sample	CpG site$_1$	CpG site$_2$	$\cdots$	CpG site$_m$	Class label
S_1	bv$_{11}$	bv$_{12}$	$\cdots$	bv$_{1m}$	Normal
S_2	bv$_{21}$	bv$_{22}$	$\cdots$	bv$_{2m}$	Tumoral
$\cdots$	$\cdots$	$\cdots$	$\cdots$	$\cdots$	$\cdots$
S_n	bv$_{n1}$	bv$_{n2}$	$\cdots$	bv$_{nm}$	Normal

2.1 Identifying HD Levels

In hyperdimensional computing, every atomic element is associated to a D-dimensional vector. In general in a dataset, the observation values can be highly

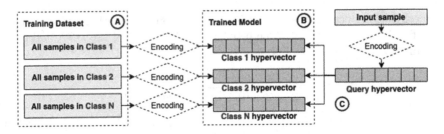

Fig. 1. Flow chart describing both the HD training and classification procedures. The training samples are grouped according to their class and used to build the HD representations of the classes by encoding the observations (A and B) of the samples. The encoding function requires to know the number of HD levels which must be chosen a priori. Thus, the HD model is used to assign a class to the HD representation of an input sample during the classification phase by selecting the most similar class hypervector (C). The input samples must be encoded with the same encoding function applied on the training samples.

variable. Considering a dataset with single-digit after the decimal point numbers, ranging from a minimum of 0.1 to a maximum of 0.9, a D-dimensional vector for each number in this interval must be built, with a total of nine vectors. These set of HD vectors represents the data dictionary that is part of the classification model. Without it, the final classification model is useless, because it will not be possible to map a value in the dataset to a vector in the D-dimensional space. More in general, dealing with real numbers, depending on their precision, means that there exists a potentially huge set of distinct values, thus an equally sized set of distinct D-dimensional vectors have to be built. However, depending on the nature of the dataset, it is always possible to round the precision of the values of the observations. This will drastically reduce the amount of vectors required to create the data dictionary at the cost of a penalty in terms of precision. The identified number of vectors per value is called the number of HD levels L. Generally, considering $delta = (Obs_{max} - Obs_{min})/L$, the original interval can be splitted into L equal sub-intervals that start from $S_i = (delta * L_i)$, where i ranges from 0 up to $L - 1$ (i.e.: a given observation value is associated to the level L_i if it is less than S_i).

2.2 Encoding Observations

Encoding data is the most crucial step. The proposed approach is able to consider how each single observation value and its position in the dataset will influence the informative content of the final observation encoded vector. This procedure is described in in Algorithm 1. The same procedure is applied to every observation in the dataset.

Algorithm 1. Encode observation

1: $vector \leftarrow [0 \dots 0]$
2: **for** $i = 0, 1, \dots, M-1$ in Obs **do**
3:     $l \leftarrow$ get the level vector according to the Obs value in the i-th position
4:     Rotate l by i positions
5:     Element-wise sum l and $vector$
6: **end for**

2.3 Training

Before the training phase, the HD vector representation of a class is composed of the sum of all the class observation HD vectors. New information are constantly added over the previous class HD representation. At the end, the final HD vector will look very different from every single class observation. The similarity between one of the class observation vectors and the final vector representation of the class is thus compromised, and an observation could results close to the wrong class. For this reason, the informative content of the HD vectors of the classes must be fixed. Here we introduced the core of the training phase in which every observation vector is compared with the HD vectors of the classes. By computing the inner product it is possible to predict a class by choosing the most similar one. If a wrong prediction occurs, the observation vector will be subtracted from the wrongly predicted class vector and added to the correct one. The training phase can be repeated multiple times until the W/O ratio will converge to zero or will remain stable, where W is the number of wrongly classified instances and O is the total number of observations. The fixed classes HD vectors, along with the level vectors, finally represent the classification model.

2.4 Classifying Dataset

Once the classification model is built, the classification of a dataset will be easy and fast. Two simple steps are needed: for each observation in the dataset: (i) represent it as HD vector following the procedure previously described exploiting the HD levels (that are part of the classification model), and (ii) predict a class by choosing the closest one according to the inner product applied on the observation vector and all the HD classes representations.

3 Results

We tested the proposed approach on three datasets containing DNA methylation values of patients affected by three different types of cancer, i.e., BRCA, KIRP, and THCA, retrieved from the public FTP repository of TCGA2BED [3]. TCGA2BED contains TCGA data converted in BED format that follow strict standardisation rules. This allowed us to easily preprocess them by creating a data matrix for each tumor that contains the methylated sites on the columns, which are the features, and the samples on the rows as shown in Table 1.

The number of samples and features are reported in Table 2. First, we randomly sample each dataset to compose the training and test sub-datasets following an 80% - 20% proportion on both tumoral and normal observations. Thus, we start encoding and training all the three datasets (80% of the observations) by using one thousand HD levels, producing three classification models, one per dataset. Finally, we start the classifier with the produced models and HD levels on the remaining 20% of the samples. The classification performances in terms of computation time and accuracy are very promising (see Table 3). The same procedure has been applied ten times to cross validate the results, producing similar results.

Table 2. Compact overview of the datasets

Dataset	Tumoral samples	Normal samples	Features
BRCA	799	98	485,512
KIRP	276	45	485,512
THCA	515	56	485,512

Table 3. Performance of the HD classification algorithm

Dataset	Training time (hours)	Classification time (seconds)	Accuracy (percentage)
BRCA	5.44	3	97.7
KIRP	2.97	1	98.4
THCA	3.68	2	100.0

We compared the achieved results with the performance of BIGBIOCL [2], a supervised classification algorithm for distributed environments, which is able to extract multiple models by performing hundreds of classification iterations on a massive amount of features in few hours. BIGBIOCL is able to extract alternative models by recomputing the classification procedure multiple times excluding the features resulted significantly relevant in the previous iterations. We applied our approach (the HD classifier) on the same datasets used for testing the performance of BIGBIOCL. In [2] the authors declared 4.21, 2.13, and 3.33 h of overall execution time for classifying the BRCA, KIRP, and THCA dataset respectively with an accuracy of 98% for BRCA and 97% for both KIRP and THCA. The software has been executed on 1 out of 66 nodes of one of the Cineca clusters with 7 threads of a 2.5 GHz Intel Xeon E5 2670 v2 CPU (20 cores) and 18 out of 128 GB of RAM. It is worth noting that Cineca is the Italian inter-university consortium and the largest Italian computing centre, while the experimentation of the HD classifier has been run on a consumer laptop with a 1.2 GHz (Intel Core M) CPU and 8 GB of RAM in single thread, highlighting that HD computing applications do not require massive computational resources.

Despite the limited amount of computational resources allocated for testing the HD classifier, our approach outperformed BIGBIOCL in terms of computational time and required amount of memory. The obtained classification accuracy is greater for the KIRP (98.4% vs 97%) and THCA (100% vs 97%) dataset, whereas comparable for the BRCA dataset (97.7% vs 98%). In Table 4 we summarize the comparison.

Table 4. Comparison of HD classifier vs BigBioCl classifier

Dataset	Time HD (hours)	Time BigBioCl (hours)	Accuracy HD (percentage)	Accuracy BigBioCl (percentage)
BRCA	5.47	4.21	97.7	98.0
KIRP	2.98	2.13	98.4	97.0
THCA	3.70	3.33	100.0	97.0

4 Discussion and Conclusion

In this work, we applied energy efficient hyperdimensional computing to classify tumoral and non-tumoral samples by analyzing their DNA methylation levels, demonstrating that the HD approach represents a valid high-efficient alternative to well-known classification methods. HD is as easy as fast, it involves two arithmetic operations only (addition and subtraction) and the inner product to compute the similarity between vectors. Additionally, the way with which data are modeled and the limited set of arithmetic operations, make the HD classifier ideal for the emergent class of new processors oriented to drastically reduce the energy consumption. Additionally, the incredible speed on comparing new observations against a predefined classification model, makes the proposed approach a perfect candidate for new near-real-time healthcare data analytics applications. In future we plan to extend the experimentation on other DNA methylation datasets in order to confirm the validity of our approach. Finally, we are investigating the possibility to use the vector representation of HD computing to improve the feature selection step for classification.

References

1. Cappelli, E., Felici, G., Weitschek, E.: Combining DNA methylation and RNA sequencing data of cancer for supervised knowledge extraction. BioData Min. **11**(1), 22 (2018)
2. Celli, F., Cumbo, F., Weitschek, E.: Classification of large DNA methylation datasets for identifying cancer drivers. Big Data Res. **13**, 21–28 (2018)

3. Cumbo, F., Fiscon, G., Ceri, S., Masseroli, M., Weitschek, E.: TCGA2BED: extracting, extending, integrating, and querying the cancer genome atlas. BMC Bioinformatics **18**(1), 6 (2017). https://doi.org/10.1186/s12859-016-1419-5

4. Imani, M., Huang, C., Kong, D., Rosing, T.: Hierarchical hyperdimensional computing for energy efficient classification. In: 2018 55th ACM/ESDA/IEEE Design Automation Conference (DAC), pp. 1–6. IEEE (2018)

5. Imani, M., Kong, D., Rahimi, A., Rosing, T.: VoiceHD: hyperdimensional computing for efficient speech recognition. In: 2017 IEEE International Conference on Rebooting Computing (ICRC), pp. 1–8. IEEE (2017)

6. Jensen, M.A., Ferretti, V., Grossman, R.L., Staudt, L.M.: The NCI Genomic Data Commons as an engine for precision medicine. Blood **130**(4), 453–459 (2017)

7. Kanerva, P.: Hyperdimensional computing: an introduction to computing in distributed representation with high-dimensional random vectors. Cogn. Comput. **1**(2), 139–159 (2009). https://doi.org/10.1007/s12559-009-9009-8

8. Luo, J., Wu, M., Gopukumar, D., Zhao, Y.: Big data application in biomedical research and health care: a literature review. Biomed. Inform. Insights **8**, BII-S31559 (2016)

9. Rahimi, A., Kanerva, P., Rabaey, J.M.: A robust and energy-efficient classifier using brain-inspired hyperdimensional computing. In: Proceedings of the 2016 International Symposium on Low Power Electronics and Design, pp. 64–69 (2016)

10. Soto, J., Rodriguez-Antolin, C., Vallespin, E., De Castro Carpeno, J., De Caceres, I.I.: The impact of next-generation sequencing on the DNA methylation-based translational cancer research. Transl. Res. **169**, 1–18 (2016)

11. Wadapurkar, R.M., Vyas, R.: Computational analysis of next generation sequencing data and its applications in clinical oncology. Inform. Med. Unlocked **11**, 75–82 (2018)

12. Weinstein, J.N., et al.: The cancer genome atlas pan-cancer analysis project. Nat. Genet. **45**(10), 1113 (2013)

13. Weitschek, E., Cumbo, F., Cappelli, E., Felici, G.: Genomic data integration: a case study on next generation sequencing of cancer. In: 2016 27th International Workshop on Database and Expert Systems Applications (DEXA), pp. 49–53. IEEE (2016)

14. Weitschek, E., Di Lauro, S., Cappelli, E., Bertolazzi, P., Felici, G.: CamurWeb: a classification software and a large knowledge base for gene expression data of cancer. BMC Bioinformatics **19**(10), 245 (2018). https://doi.org/10.1186/s12859-018-2299-7

TopicsRanksDC: Distance-Based Topic Ranking Applied on Two-Class Data

Malik Yousef[1,2(✉)], Jamal Al Qundus[3], Silvio Peikert[3], and Adrian Paschke[3]

[1] Zefat Academic College, Zefat, Israel
malik.yousef@gmail.com
[2] The Galilee Digital Health Research Center (GDH), Zefat, Israel
[3] Data Analytics Center (DANA), Fraunhofer FOKUS, Berlin, Germany
{jamal.al.qundus,silvio.peikert,adrian.paschke}@fokus.fraunhofer.de

Abstract. In this paper, we introduce a novel approach named TopicsRanksDC for topics ranking based on the distance between two clusters that are generated by each topic. We assume that our data consists of text documents that are associated with two-classes. Our approach ranks each topic contained in these text documents by its significance for separating the two-classes. Firstly, the algorithm detects topics using Latent Dirichlet Allocation (LDA). The words defining each topic are represented as two clusters, where each one is associated with one of the classes. We compute four distance metrics, Single Linkage, Complete Linkage, Average Linkage and distance between the centroid. We compare the results of LDA topics and random topics. The results show that the rank for LDA topics is much higher than random topics. The results of TopicsRanksDC tool are promising for future work to enable search engines to suggest related topics.

Keywords: Topic ranking · Clusters distance · Cluster significant

1 Introduction

Information is regularly digitally generated and archived. Searching for documents in digital archives becomes more and more difficult over time, and parallel to this, the number of use cases and their interrelationships that need to be covered by information retrieval systems are growing. In this sense, there is a constant need for techniques that help us to organize, explore and understand data. One important technique is topic modeling that performs analysis on texts to identify topics. These topic models are used to classify documents and to support further algorithms to perform context adaptive feature, fact and relation extraction [1]. While Latent Dirichlet Allocation (LDA)[2], Pachinko Allocation [3], or Probabilistic Latent Semantic Analysis (PLSA) [4] traditionally perform

This work has been partially supported by the "Wachstumskern Qurator – Corporate Smart Insights" project (03WKDA1F) funded by the German Federal Ministry of Education and Research (BMBF).

© Springer Nature Switzerland AG 2020
G. Kotsis et al. (Eds.): DEXA 2020 Workshops, CCIS 1285, pp. 11–21, 2020.
https://doi.org/10.1007/978-3-030-59028-4_2

topic modeling by statistical analysis of co-occurring words, the approaches [5] [6] in integrate semantics into LDA. In general, topic modeling is about capturing the relationships among the words and documents (in terms of topics), and calculating the likelihood of belonging a word to a topic. However, in many situations when working with large amounts of data, it is very difficult to help the user to quickly grasp a text collection or to get an overview of identified topics. Therefore, it is helpful to rank all topics and focus on the relevant ones that are both pressing and significant.

This is precisely the aim of topic ranking approaches, namely not to offer all identified topics in the same way, but to investigate the correlation of topics that correspond to a given sector and to present those with higher priority. Ranking of topics could lead to a loss of information, which occurs when relevant topics or their clusters are ranked too low and are therefore no longer represented in the ranking. This problem can be alleviated by merging corresponding or similar clusters. This leads to the research question: To what extent can measuring the cluster distance of topics support the topic ranking? The intuitive and basic principle for calculating distances fits the Euclidean equation: $d_{euclidean}(x, y) = \sqrt{\sum_{i=1}^{n}(y_i - x_i)^2}$. The challenge is to select the appropriate representative points of the clusters. This selection depends on the desired technique for measuring distance. In the case of a single-linkage, these points are the closest between the two clusters. In contrast, Complete-link distance uses the points furthest from the other cluster. While the average-link applies the average distance between all points of both clusters, otherwise, the distance of the centroids the clusters can be considered. This paper aims to determine the degree of likeness of two clusters, putting the focus on the minimum distance between the clusters and thus make use of the single- linkage method. The paper is structured as follows: Sect. 2 provides a brief overview of relevant works. Section 3 describes. Section 3 presents the preparation. Section 4 reports and discusses the findings. Section 5 contains the summary and concludes with proposals for further investigation.

2   Related Work

Topic Modeling and Ranking are very popular and in great demand. Several previous studies have investigated these topics and their development, which is also being followed intensively by industry and scientists. The methods used are basically comparable, but differ in terms of the objectives, such as reducing the number of insignificant, similar or even widely divergent topics. It should be noted that summary does not always have to be the result of a topic-relevant search, as topics can never match 100%. The study of [7] presents a semantic-based approach for the automatic classification of LDA topics to finally identify junk and insignificant topics. The authors measure how "different" the distribution of topics is as a "junk" distribution and thus the degree of insignificance a derived topic carries in its distribution. According to the Zipf law[1]: A real topic can be modelled as a distribution of small number of words, so-called "salient words". Conversely, the "junk distribution" means that a large number of terms probably represent insignificant topics. Kullback-Leibler (KL)[2] divergence was used to calculate the distance of the topic distribution over the number of salient words. Unlike our work, this work is based on an unsupervised quantification of topic meaning by identifying junk and insignificant topics. While our work aims to find relevant topics. The study of [8] proposes a method for re-ranking topics, motivated by the observation that similar documents are likely to be related to similar topics, while their diversity indicates the likely association with different topics. They developed the metric Laplacian score to rank topics, reflecting the degree of its discriminatory documents, in order to find the topics with high levels of discrimination, as well as the paired mutual information that calculates the information that two topics have in common. In this way, the similarity of topics can be measured, which can be used to maximize the diversity of topics and minimize redundancy. With the similar aim of measuring how different the topics are, [9] applied the "Non-Markov Continuous-Time Model of Topical Trends" to calculate the average distance of word distributions between all topic pairs. As the authors themselves claimed, this method of calculating topic similarity is better suited to reduce topic redundancy. The study of [10] proposed a method for evaluating the quality of each topic, using the metric of the silhouette coefficient. Using the latent topic model (LDA), the approach is based on clustering topics and using the silhouette index, which is often used to characterize the quality of elements in a cluster. Topics from multiple models are clustered based on the similarity of their word distributions. The quality of learned clusters is examined to distinguish weak topics from strong topics. In this work the clusters (e.g. weak and strong) are left unchallenged without further measuring the correlation of these clusters e.g. to investigate Euclidean distances in the metrics.

3 Methodology

3.1 Data

We have considered different two-class text data sets. The first data set consists of short texts downloaded from the repository Stack Overflow[3]. Applying the trust model proposed by [11,12] the data is classified into four classes: *very-trusted* (844 entries), *very-untrusted* (117 entries), *trusted* (904 entries) and *untrusted* (347 entries). As we consider two-class data, we divided the data into the binary

---

[1] Joachims, T.: A Statistical Learning Model of Text Classification with Support Vector Machines. In: Proceedings of the Conference on Research and Development in Information Retrieval, SIGIR (2001).

[2] Bishop, C.M.: Pattern Recognition and Machine Learning. Springer, Heidelberg (2006).

[3] https://archive.org/details/stackexchange.

set *trusted/untrusted*, and *very-trusted/very-untrusted* documents. The second data set is created from two sources -Human-Aids and Mouse Cancer with 150 instances downloaded from PubMed. For simplicity, we will refer to this data as Aids vs Cancer. A pre-processing procedure is applied on the data in order to transform it into a vector space data that could be subject to our algorithm.

3.2   Pre-processing

A pre-processing step is necessary to convert the texts into vector space representations. We have used Knime [14] workflows for text preprocessing. Firstly, we perform cleaning of the text data using the following procedure: Punctuation Erasure, N-chars Filter, Number Filter, Case Converter (lower case), Stop-words Filter, Snowball Stemmer and Term Filtering. We used a language detector provided by Tika-collection to process English texts only. In the next step, the friction words are used as a dictionary that represents each document. These dictionaries are called bag-of-words (BoW) representation. BoW can be represented by Term-Frequency (TF) or binary. In the TF format counts the number of times a word appears in the document while the binary representation only distinguishes between 1 if the word is present and 0 otherwise. The number of words/features after performing the pre-processing stage is 714 for trusted vs untrusted data set, 848 words for the very-trusted vs very-untrusted data set and 1440 words for the human-aids vs mouse-cancer data set. For more detail see [13].

3.3   Topic Clustering

The distance metric for clusters is a very important parameter for different clustering approaches. We use an agglomerative approach to hierarchical clustering. During agglomerative clustering, we merge the closest clusters as defined by the distance metric chosen. For computing the distance of two clusters, there is a variety of possible metrics, in this study we will use the four most popular: single-linkage, complete-linkage, average-linkage, and centroid-linkage.

Single-link distance defines the distance between two clusters as the minimum distance between their members or is the distance between the nearest neighbors: $d_1(c_1, c_2) = \min_{x \in c_1, y \in c_2} ||x - y||$.

It is called "single link" because it defines clusters that are close, if they have even a single pair of close points. Complete-link distance is defined as the distance between clusters. It is the maximum distance between the points of the clusters, or it is the distance between farthest neighbors: $d_2(c_1, c_2) = \max_{x \in c_1, y \in c_2} ||x - y||$.

Average-linkage is the distance between each pair of points. In each cluster those are added up and divided by the number of pairs, which results in average inter-cluster distance. $d_3(c_1, c_2) = \frac{1}{n_{c_1} n_{c_2}} \sum_{i=1}^{n_{c_1}} \sum_{j=1}^{n_{c_2}} d(x_{i_{c_1}}, x_{j_{c_2}})$ Centroid-linkage is the distance between the centroids of two clusters. $d_4(c_1, c_2) = ||\bar{x_{c_1}} - \bar{x_{c_2}}||$, where $\bar{x_{c_1}}$ is the centroid for the cluster c_1, while $\bar{x_{c_2}}$ is the centroid for cluster c_2.

3.4   TopicsRanksDC Algorithm

As illustrated in Fig. 1, the input to the algorithm is a collection of text documents that we assume to be of two classes. We mean by two-class data where part of the examples belongs to one label while the other belongs to the second label (documents about cancer patients vs documents about healthy). The algorithm consists of two stages; the first stage detects the topics using LDA using the whole data (all the collection of text). A topic is a bag of words. The second stage is ranking or scoring each topic based on the distance of two clusters, where each cluster represents examples belongs to one label based only on the words of the given topic (Fig. 1). The ranks stage actually defines the significance of each word in a topic in terms of separating the two-classes. In other words, suppose one has thousands of words that represent some data, and one wants to find out which groups of words (topics) were suitable to separate the two-classes. Let assume that we have detected n topics. Each topic contains k words. In order to calculate the rank or the significance of each topic related to the two classes we perform the following algorithm:

For each topic i (i=1,..,n) perform:
    a)   Create two clusters c1 and c2 of points that are represented by the words belonging to topic i. c1 contains the points of the first class (positive) and c2 contains the points of the second class (negative).
    b)   Calculate four distance metrics, Single Linkage, Complete Linkage, Average Linkage and distance between the two centers so-called ''centroids''.

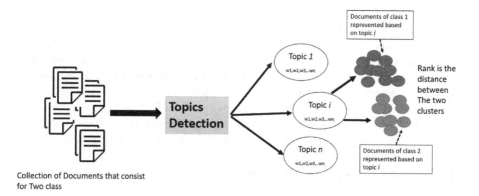

Fig. 1. The main workflow of TopicsRanksDC tool. The input is a collection of text documents belongs to two-class. Next is LDA or other approach for detecting topics. The last stage is ranking/scoring the topics on two-class data by distance of two clusters.

The rank we calculate gives an indication of how important the topic is for the separation of the two given classes, considering only the words associated with the specific topic. If the rank is close to zero, this implies that the two clusters are inseparable and the topic is not important to distinguish the two classes.

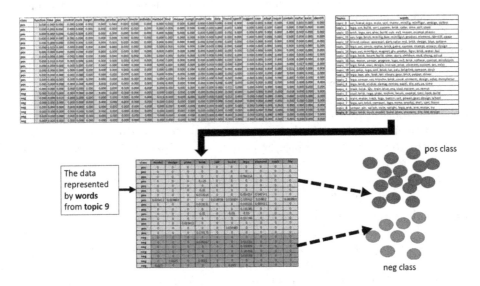

Fig. 2. Illustrates of how the data is represented by words that belong to a specific topic creating new data with just columns that are associated with the words. Rows are the sample. The new data appears in the lower side of the figure is the two clusters.

The data in Fig. 2, consists of columns that contain the features/words, while the rows are associated with documents. The columns class represents the labels for each document. In our case we assume two-class data sets (positive vs negatives). The new data appears on the lower part of the figure and is represented by the words belonging to topic number 9. The topics and their word lists appear on the upper part on the right side. The two clouds or clusters represent the new data in two- dimensional space. The aim is to calculate the distance between these two clusters. If the clusters are separable then the score/rank should be high, while if the two clusters are non- separable, the score should be close to zero indicating that the topic words are insignificant for the two-classes.

4   Results and Discussion

In order to have an idea about the two-classes, we have evaluated the performance of Random Forest (RF) on each data. The classifiers were trained and tested, with the division into 90% training data and 10% test data from the data generated by the pre-processing phase. The trusted/untrusted data sets used by the classifier are imbalanced, which can influence the classifier to the advantage of the set with more samples and is so called the problem of the imbalanced class distribution. We have applied an under-sampling approach that reduces the number of samples of the majority class to the minority class, thus reducing the bias in the size distribution of the data subsets. For more details see [13]. We have set the ratio of the reduction to be up to 2 fold. We applied 100-fold Monte Carlo Cross Validation (MCCV)[15]. The average of the 100 iteration is calculated to form as the final performance evaluation. Table 1, presents the results of RF. It is clear that the data of Aids—Cancer is much more separable than the data of trusted—untrusted. The result here is with considering all the features generated by the pre-process step to bring the raw data into vector space. In order to test the algorithm TopicsRanksDC, we set the number of topics to be 10. Then we generate different size of topics. Size is the number of words in each topic. For each one of these options, we compute the significant/ranks of each topic. Table 2 illustrates one sample of the output of the tool.

Table 1. Classifier performance of the random forest based on bag-of-words model.

Data set	Sen	Spe	F1	Acc
vt vs vu	0.95	0.72	0.84	0.76
t vs u	0.98	0.85	0.80	0.69
Aids vs Cancer	0.98	0.93	0.96	0.96

vt = very   trusted,   t = trusted,
u = untrusted,   vu = very   untrusted.
Sen is sensitivity, Spe is specificity, F1
is F1 measure and Acc is accuracy

To add sense of these values, we have generated random topics that include random words from the whole sets. The random topics used as input to the tool with the input of the algorithm binary data. The results of this experiment are illustrated in Table 3.

Table 4 presents the ratio ranks between the topics generated by LDA and the random topics. The MeanD (average-link distance or d3 metric) shows ratio of above 3.6 for all the topics. Figure 3 summarizes the results ratio of TopicsRanksDC output applied on LDA topics and random topics. The topic 10 with words 5 is considered. Generally, Fig. 3 shows that the model of TopicsRanksDC performs well and is much higher than the random topics. For example, for the Aids vs Cancer data the centroids of the two-classes represented by just the top ranked topics is very large. These results match the higher accuracy when using random forest.

We tested TopicsRanksDC on fixed number of topics with variety of number of words. The 10 topics were generated by LDA with different count of words in the topic models. We consider 5, 10, 20 and 40 words as the size of topic models.

Figure 4a and b show the influence of the size of the words (number of topics is fixed to 10). In Fig. 4 that related to the t vs u data, we removed the results for

Table 2. TopicsRnaksDC result on the data t vs. u

Cluster	Topic terms	CentroidD	MinD	MaxD	MeanD
topic_5	lego, piec, set, brick, model	0.10	0.00	2.24	1.18
topic_6	lego, set, brick, piec, box	0.08	0.00	2.24	1.16
topic_1	set, lego, http, list, bricklink	0.09	0.00	2.24	1.04
topic_7	brick, lego, stud, plate, piec	0.08	0.00	2.24	1.02
topic_0	brick, lego, motor, gear, batteri	0.09	0.00	2.24	0.91
topic_9	lego, block, water, duplo, set	0.07	0.00	2.24	0.90
topic_8	motor, lego, power, gear, us	0.09	0.00	2.24	0.88
topic_4	nxt, program, sensor, ev3, lego	0.04	0.00	2.24	0.81
topic_3	unit, pod, set, oppon, build	0.06	0.00	1.73	0.59
topic_2	friend, accessori, brick, hole, print	0.06	0.00	2.00	0.52

Table 2 shows the result of applying TopicsRnaksDC on the data trusted vs untrusted (t vs u). The number of LDA topics is 10 with 5 words in each topic. CentroidD is the Centroid-linkage, MinD is the Single-link distance, MaxD is the Complete-link distance, while MeanD is the average-link

Table 3. TopicsRnaksDC result on words in each topic selected randomly

Cluster	Topic terms	CentroidD	MinD	MaxD	MeanD
topic_3	happen, seri, fit, tell, avail	0.02	0.00	2.00	0.31
topic_9	get, prefer, fan, question, file	0.03	0.00	2.00	0.29
topic_6	tend, floor, probabl, mean, stand	0.05	0.00	2.24	0.29
topic_7	usual, easi, etc, structur, heavi	0.05	0.00	2.00	0.25
topic_8	half, take, stack, coupl, recommend	0.04	0.00	1.73	0.25
topic_4	stick, seller, steer, mindstorm, suspect	0.06	0.00	2.00	0.22
topic_1	war, car, guid, access, stop	0.02	0.00	1.73	0.17
topic_2	speed, calcul, shape, addit, save	0.02	0.00	2.00	0.16
topic_5	assembli, techniqu, regard, environ, displai	0.01	0.00	2.00	0.14
topic_0	place, true, imagin, vertic, particular	0.02	0.00	1.73	0.13

Table 3 represents the result of applying TopicsRnaksDC on the data trusted vs untrusted (t vs u.) The number of topics is 10 with 5 words in each topic selected randomly form all the words.

MinD as all zeros, while it is clear that MaxD metrics are increasing significantly as the number of words is increasing and a slight increase is in the MeanD metric. The CentroidD metric is almost similar. Similar observations for the MaxD metric are presented in Fig. 4b for the Aids vs Cancer data. Additionally, the MinD is getting increase indicating that the data is more separable than the t vs u data. Also, the MeanD metric is increasing significantly.

Table 4. Ratio ranks between the topics generated by LDA and the random topics.

CentroidD	MaxD	MeanD
5.92	1.12	3.78
2.80	1.12	4.01
1.71	1.00	3.59
1.54	1.12	3.99
2.13	1.29	3.66
1.08	1.12	4.17
4.25	1.29	5.03
1.61	1.12	4.96
5.25	0.87	4.23
2.76	1.15	4.05

Table 4 shows the ratio between values of LDA to random topics. The MinD column is discared is all is zero for all topics.

Ratio with Random Topics

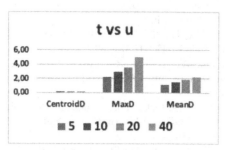

Fig. 3. Destruction of ratio with the random topics.

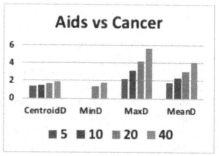

(a) The distribution of the 3 distance metrics with different number of words in 10 topics. We remove the MinD as it zero for all words. The data is t vs u .

(b) The distribution of the 4 distance metrics with different number of words in 10 topics. The data Aids vs Cancer.

Fig. 4. The distribution of distance metrics over two datasets.

5   Conclusion and Future Work

This work introduced a novel approach for topic ranking or scoring applied on two-class data sets. The score is actually the significance of the topic (set of words) for separating the two-classes. The scoring function used for computing the ranks is based on the distance between clusters associated with one of the classes represented by the words belong to a specific topic. Interestingly also simple metrics were successful in ranking the topics comparing to random topics. However, we assume that more metrics should be examined and as a future work, we will consider to use the machine learning approaches for ranking topics. One approach is considering the one-class classifier applied on text data [16,17].

References

1. Al Qundus, J., Peikert, S., Paschke, A.: AI supported topic modeling using KNIME-workflows. In: Conference on Digit Curation Technologies, Berlin, Germany (2020)
2. Blei, D.M., Ng, A.Y., Jordan, M.I.: Latent Dirichlet allocation. J. Mach. Learn. Res. **3**, 993–1022 (2003)
3. Wei, L., McCallum, A.: Pachinko: allocation DAG-structured mixture models of topic correlations. In: ACM International Conference Proceeding Series (2006)
4. Hofmann, T.: Probabilistic latent semantic indexing. In: Proceedings of the 22nd Annual International ACM SIGIR Conference on Research and Development in Information Retrieval, SIGIR 1999 (1999)
5. Allahyari, M., Kochut, K.: Automatic topic labeling using ontology-based topic models. In: Proceedings - 2015 IEEE 14th International Conference on Machine Learning and Applications, ICMLA 2015 (2016)
6. Hulpus, I., Hayes, C., Karnstedt, M., Greene, D.: Unsupervised graph-based topic labelling using DBpedia. In: WSDM 2013 - Proceedings of the 6th ACM International Conference on Web Search Data Mining (2013)
7. AlSumait, L., Barbará, D., Gentle, J., Domeniconi, C.: Topic significance ranking of LDA generative models. In: Buntine, W., Grobelnik, M., Mladenić, D., Shawe-Taylor, J. (eds.) ECML PKDD 2009. LNCS (LNAI), vol. 5781, pp. 67–82. Springer, Heidelberg (2009). https://doi.org/10.1007/978-3-642-04180-8_22
8. Song, Y., Pan, S., Liu, S., Zhou, M.X., Qian, W.: Topic and keyword re-ranking for LDA-based topic modeling. In: International Conference on Information and Knowledge Management Proceedings (2009)
9. Wang, X., McCallum, A.: Topics over time: a non-Markov continuous-time model of topical trends. In: Proceedings of the ACM SIGKDD International Conference on Knowledge Discovery and Data Mining (2006)
10. Mehta, V., Caceres, R.S., Carter, K.M.: Evaluating topic quality using model clustering. In: IEEE SSCI 2014–2014 IEEE Symposium on Computational Intelligence and Data Mining, Proceedings (2015)
11. Al Qundus, J., Paschke, A., Kumar, S., Gupta, S.: Calculating trust in domain analysis: theoretical trust model. Int. J. Inf. Manage. **48**, 1–11 (2019)
12. Qundus, J.A., Paschke, A.: Investigating the effect of attributes on user trust in social media. In: Elloumi, M., Granitzer, M., Hameurlain, A., Seifert, C., Stein, B., Tjoa, A.M., Wagner, R. (eds.) DEXA 2018. CCIS, vol. 903, pp. 278–288. Springer, Cham (2018). https://doi.org/10.1007/978-3-319-99133-7_23

13. Al Qundus, J., Paschke, A., Gupta, S., Alzouby, A., Yousef, M.: Exploring the impact of short text complexity and structure on its quality in social media. J. Enterp. Inf. Manage. (2020)
14. Berthold, M.R., Cebron, N., Dill, F., Gabriel, T.R., Kötter, T., Meinl, T., et al.: KNIME: the Konstanz information miner. SIGKDD Explor. 319–326 (2008)
15. Xu, Q.-S., Liang, Y.-Z.: Monte Carlo cross validation. Chemom. Intell. Lab. Syst. **56**, 1–11 (2001)
16. Manevitz, L., Yousef, M.: One-class document classification via Neural Networks. Neurocomputing **70**, 1466–81 (2007)
17. Manevitz, L.M., Yousef, M.: One-class SVMs for document classification. J. Mach. Learn. Res. **2**, 139–154 (2001)

Cyber-Security and Functional Safety in Cyber-Physical Systems

YASSi: Yet Another Symbolic Simulator Large (Tool Demo)

Sebastian Pointner[1]([⊠]), Pablo Gonzalez-de-Aledo[1], and Robert Wille[1,2]

[1] Johannes Kepler University Linz, Linz, Austria
{sebastian.pointner,robert.wille}@jku.at, pablo.aledo@gmail.com
[2] Software Competence Center Hagenberg GmbH (SCCH), Hagenberg, Austria

Abstract. Safety critical systems have finally made their way into our daily life. While recent industrial and academic research could already improve the design cycle for such systems, ensuring the functionality of such systems still remains an open question. Such systems which are composed of hardware as well as software components have to be checked since any wrong behavior of the system could end up in harming human life. To this end, program analysis techniques can be applied in order to ensure that the program works as intended and that no unwanted behavior is executed. However, approaches like static or dynamic program analysis which are widely applied for this purpose still lead a large number of fault positive results. To overcome such limitations an alternative approach called symbolic execution has been proposed. In this work, we present a tool called YASSi which implements this approach. Applying YASSi allows to symbolically execute programs written in the C/C++ language. By this, YASSi can be applied for several applications needed for the checking program for safety critical properties like (1) assertion checking, (2) reachability analysis, or (3) stimuli generation for digital circuits.

Keywords: Symbolic simulation · Assertion checking · Stimuli generation

1 Introduction

The technical progress achieved by academia as well as industry within the last decades led to more and more complex systems. These systems, which are composed of software and hardware components, have made their way into our daily life and are especially very important in terms of functional safety applications. To this end, it is of utter most importance to ensure that applications like the trigger unit of an airbag or a breathing apparatus for medial emergencies work as intended.

In order to ensure that such systems are getting realized correctly, the hardware as well as the software part of the system has to be checked for correctness.

G. Kotsis et al. (Eds.): DEXA 2020 Workshops, CCIS 1285, pp. 25–31, 2020.
https://doi.org/10.1007/978-3-030-59028-4_3

Since the hardware and the software part of the system are getting designed with abstract programming languages, program analysis techniques can be applied for this purpose. To this end, static program analysis like Control-Flow-Analysis [1] as well as dynamic program analysis like Dynamic-Program-Slicing [2] have emerged in the past. However, static as well as dynamic program analysis techniques both have major limitations (e.g. a significant number of false positives). Hence, in contrast to these program analysis techniques, the approach of symbolic execution emerged in the past [3]. Symbolic execution has been investigated heavily in the past and already led to a significant number of tools. These tools— including KLEE [4], DIVINE [5], Forest [6], or CBMC [7]—allow to symbolically execute programming languages like C/C++. However, symbolic execution is not limited to C/C++ only. Approaches like [8] also show that those methods can even be utilized for the execution of abstract hardware descriptions such as provided in SystemC.

In this work, we present the tool YASSi (*Yet Another Symbolic Simulator*) as our state-of-the-art approach for the realization of symbolic execution for academic research. Our approach is capable to symbolically execute program code written in the C/C++ language. To this end, YASSi is based on the *Low Level Virtual Machine* (LLVM) compiler infrastructure [9] and utilizes the power of modern reasoning engines like Z3 [10] or Boolector [11] for decision finding. Applying YASSi allows the user to symbolically explore his/her design and, therefore, to check certain design properties or to perform reachability analysis. Compared to other state-of-the-art symbolic execution tools, we intentionally have designed YASSi in a fashion that it is easily extendable for academic research purposes.

The remainder of this paper is structured as follows. The next section briefly reviews the principle of symbolic execution for program analysis. Based on this, we are introducing YASSi as our approach for an academic tool for symbolic execution in Sect. 3. Afterwards, we discuss the applicability of YASSi for specific applications in Sect. 4, before we conclude the paper in Sect. 5.

2   Background

This section briefly reviews the approach of symbolic execution as an advanced program analysis technique [3]. In general, symbolic execution rests on the idea of rather than using concrete values for so-called *free variables* (e.g. the inputs of a function), the value of such variables is treated symbolically.

Example 1. *Figure 1a shows a code snippet composed of using two free variables and three conditional statements. Depending on the value held by the variables variable_a and variable_b, the execution of the code snippet returns a different value back to the host system.*

In order to perform a symbolic execution, the system has to keep track of all possible execution possibilities. To this end, symbolic execution has to evaluate all possible outcomes of branches within the source code. Moreover, the execution engine keeps track of the positive branches (i.e. the taken branch) as well as the negative branches (i.e. the non-taken branch).

Example 2. *Consider again Fig. 1b which shows the Control Flow Graph (CFG) of the code snippet. Every time the execution reaches a branching condition, it is checked whether there is a possible valid assignment for the branching variable. In case of the branch as shown on top of Fig. 1b, the symbolic execution engine has to ensure that there is a valid assignment possible for variable_a which ends up in taking the branch as well as not taking the branch.*

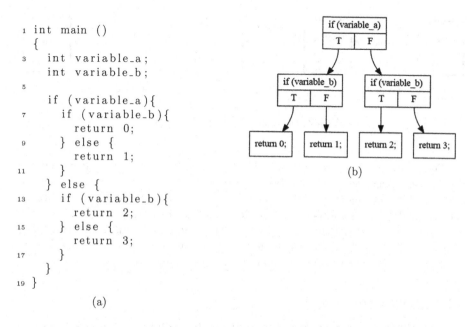

```
1  int  main  ()
   {
3     int  variable_a;
      int  variable_b;
5
      if  (variable_a){
7        if  (variable_b){
           return  0;
9        } else {
           return  1;
11       }
      } else {
13       if  (variable_b){
           return  2;
15       } else {
           return  3;
17       }
      }
19 }
```

(a)

(b)

Fig. 1. Source code and according CFG showing nested branches.

The symbolic execution engine has to decide whether a branch can be taken or not. To this end, modern symbolic execution engines are invoking the power of modern reasoning engines, e.g. for *Satisfiability Modulo Theories* (SMT), for this purpose. Moreover, the so-called *branching conditions* can be formulated using the so-called SMT2 constraint language [12]. The reasoning engine decides for every branching condition if there is an assignment possible for every *free variable* in order to take or not take the branch. The solutions generated by the reasoning engine are then directly applied by the symbolic execution engine to perform the execution of the target code.

3   The YASSi Tool

This section now introduces the YASSi symbolic simulation tool[1]. To this end, the section first introduces the modular architecture of YASSi in general before the major components of YASSi (i.e. the front-end and the back-ends) are getting discussed.

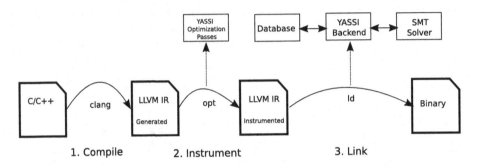

Fig. 2. Basic architecture of YASSi.

The basic architecture of YASSi is illustrated in Fig. 2. As can be seen in the figure, YASSi performs several steps before the symbolic execution can be performed by running a special binary. To this end, the generation of this binary is controlled YASSi's front-end. This binary is linked to particular YASSi back-end which keep track of the symbolic execution and commands over a reasoning engine for decisions making. In the following, we are now describing the YASSi front-end as well as the YASSi symbolic execution core back-end.

3.1   YASSi's Frontend

In the architecture of YASSi, we are using the front-end in order to prepare the program we want to execute symbolically. As can be seen in Fig. 2, the first step is the compilation of C/C++ sources into LLVM's *Intermediate Representation* (IR) format [13]. LLVM's IR format is based on a load and store architecture and breaks down the complexity of C/C++ codes into a basic instruction set. In the second step, the generated IR code is getting instrumented using LLVM's optimization tool. To this end, we are going to alternate LLVM's IR code and add special function calls for each particular instruction which is needed for the eventual symbolic execution. Once the code instrumentation has been done, we are linking the instrumented code to our back end which resolves the inserted function calls. The result of YASSi's front-end is a binary which is ready for symbolic execution. After the execution of the binary has terminated, the front-end can access the database in order to analyze the results generated by the symbolic execution run.

---

[1] YASSi is available at http://github.com/gledr/YASSi.

3.2 YASSi's Backends

YASSi commands over multiple back-ends. However, in this work we are only introducing YASSi's symbolic execution core back-end. As introduced above, YASSi's front-end performs code instrumentation by inserting callback function calls which are getting resolved by the back-end. Instead of executing the LLVM instructions, we are calling YASSi's back-end which processes the information internally. To this end, YASSi is controlling the program execution based on Z3 which is getting used as reasoning engine in the back. Moreover, every-time the program branches, YASSi considers both execution paths and tries to find a valid solution for both paths. Therefore, YASSi creates clauses for each path-condition and forwards them to the reasoning engine. If the reasoning engine determines a valid solution, the branch is getting considered successfully and the execution continues. Same as for branches is getting used for assertions as well as other exceptions.

4 Application of YASSi

This section finally discusses some of the applications YASSi can be used for. Since the principle of symbolic execution allows it to target a wide variety of applications, we focus on those which we successfully applied with YASSi thus far.

- Assertion Checking:
  YASSi is capable to check certain properties during execution. To this end, YASSi is capable to check assertion and to work with non-deterministic variables. Therefore, YASSi tries to violate the assertion by invoking the SMT solver. YASSi is not only capable to check assertions, it is also capable to check for traps like division by zero or out of boundary index accessing of data-structures like arrays.
- Reachability Analysis:
  Another application case for YASSi which has been applied successfully is reachability analysis. To this end, we have applied YASSi in order to check, that certain parts of the code are unreachable which we could use to exclude certain functional safety issues.
- Stimuli Generation:
  Next to exception checking and reachability analysis, we were able to use YASSi for stimuli generation. To this end, we were able to generate stimuli with a coverage for abstract descriptions of digital circuits. Our approach allowed it to extract these stimuli into a database and directly to apply them for checking the model. As already mentioned above, YASSi commands over multiple back-ends. Moreover, YASSi commands over a so-called replay back-end which directly can be applied to check the reached branch and line coverage [14] for a particular set of generated stimuli.

5   Conclusion

In this work, we considered symbolic simulation as program analysis technique for C/C++ codes. This is motivated by the ever growing complexity of modern systems build up using hardware as well as software components designed using programming languages like C/C++. In order to address this, we introduced the tool YASSi as our approach for a state-of-the-art symbolic simulator for academic research. We have build YASSi in a modular fashion, which allows it directly to extend the tool for eventual later applications. YASSi has been build on top of the LLVM toolkit for compiler construction together with modern reasoning engines like Z3. YASSi is further under heavy development, and we keep adding more applications. The next milestone for the tool will be the support of floating-point variable based on the SMT2 bitvector floating-point type. YASSi is available at http://github.com/gledr/YASSi.

Acknowledgments. This work has partially been supported by the LIT Secure and Correct Systems Lab funded by the State of Upper Austria as well as by BMK, BMDW, and the State of Upper Austria in the frame of the COMET Programme managed by FFG.

References

1. Midtgaard, J.: Control-flow analysis of functional programs. ACM Comput. Surv. CSUR **44**, 1–33 (2012)
2. Agrawal, H., Horgan, J.R.: Dynamic program slicing. SIGPLAN Not. **25**, 246–256 (1990)
3. King, J.C.: Symbolic execution and program testing. Commun. ACM (1976)
4. Cadar, C., Dunbar, D., Engler, D.R.: KLEE: unassisted and automatic generation of high-coverage tests for complex systems programs. In: Proceedings of the Conference on Operating Systems Design and Implementation, San Diego, USA (2008)
5. Baranová, Z., et al.: Model checking of C and C++ with DIVINE 4. In: Automated Technology for Verification and Analysis, Pune, India (2017)
6. Gonzalez-de-Aledo, P., Sanchez, P.: FramewORk for embedded system verification. In: Baier, C., Tinelli, C. (eds.) TACAS 2015. LNCS, vol. 9035, pp. 429–431. Springer, Heidelberg (2015). https://doi.org/10.1007/978-3-662-46681-0_36
7. Clarke, E., Kroening, D., Lerda, F.: A tool for checking ANSI-C programs. In: Jensen, K., Podelski, A. (eds.) TACAS 2004. LNCS, vol. 2988, pp. 168–176. Springer, Heidelberg (2004). https://doi.org/10.1007/978-3-540-24730-2_15
8. Herdt, V., Le, H.M., Große, D., Drechsler, R.: Verifying SystemC using intermediate verification language and stateful symbolic simulation. IEEE Trans. CAD **38**, 1359–1372 (2019)
9. Lattner, C., Adve, V.: LLVM: a compilation framework for lifelong program analysis & transformation. In: Proceedings of the International Symposium on Code Generation and Optimization, San Jose, USA (2004)
10. de Moura, L., Bjørner, N.: Z3: an efficient SMT solver. In: Ramakrishnan, C.R., Rehof, J. (eds.) TACAS 2008. LNCS, vol. 4963, pp. 337–340. Springer, Heidelberg (2008). https://doi.org/10.1007/978-3-540-78800-3_24

11. Niemetz, A., Preiner, M., Biere, A.: Boolector 2.0 system description. J. Satisfiability Boolean Model. Comput. **9**, 53–58 (2015)
12. Barrett, C., Stump, A., Tinelli, C.: The SMT-LIB Standard - Version 2.0. Technical report, New York University (2010)
13. The LLVM Team: clang: a C language family frontend for LLVM. Accessed 23 Mar 2020
14. Martin, G., Bailey, B., Piziali, A.: ESL Design and Verification: A Prescription for Electronic System Level Methodology. Morgan Kaufmann Publishers Inc., San Francisco (2007)

Variational Optimization of Informational Privacy

Mohit Kumar[1,2]([✉]), David Brunner[2], Bernhard A. Moser[2],
and Bernhard Freudenthaler[2]

[1] Faculty of Computer Science and Electrical Engineering, University of Rostock,
Rostock, Germany
mohit.kumar@uni-rostock.de
[2] Software Competence Center Hagenberg, Hagenberg, Austria
{David.Brunner,Bernhard.Moser,Bernhard.Freudenthaler}@scch.at

Abstract. The datasets containing sensitive information can't be publicly shared as a privacy-risk posed by several types of attacks exists. The data perturbation approach uses a random noise adding mechanism to preserve privacy, however, results in distortion of useful data. There remains the challenge of studying and optimizing privacy-utility tradeoff especially in the case when statistical distributions of data are unknown. This study introduces a novel information theoretic framework for studying privacy-utility tradeoff suitable for multivariate data and for the cases with unknown statistical distributions. We consider an information theoretic approach of quantifying privacy-leakage by the mutual information between sensitive data and released data. At the core of privacy-preserving framework lies a variational Bayesian fuzzy model approximating the uncertain mapping between released noise added data and private data such that the model is employed for variational approximation of informational privacy. The suggested privacy-preserving framework consists of three components: 1) Optimal Noise Adding Mechanism; 2) Modeling of Uncertain Mapping Between Released Noise Added Data and Private Data; and 3) Variational Approximation of Information Privacy.

Keywords: Privacy · Information theory · Variational optimization

1 Introduction

A machine learning or a data analytics algorithm operates on datasets which might contain private data or sensitive information. The data-owner might not be

The research reported in this paper has been supported by the Austrian Research Promotion Agency (FFG) Grant 873979 "Privacy Preserving Machine Learning for Industrial Applications", EU Horizon 2020 Grant 826278 "Securing Medical Data in Smart Patient-Centric Healthcare Systems" (Serums), and the Austrian Ministry for Transport, Innovation and Technology, the Federal Ministry for Digital and Economic Affairs, and the Province of Upper Austria in the frame of the COMET center SCCH.

© Springer Nature Switzerland AG 2020
G. Kotsis et al. (Eds.): DEXA 2020 Workshops, CCIS 1285, pp. 32–47, 2020.
https://doi.org/10.1007/978-3-030-59028-4_4

willing to share data even for machine learning purposes, as a privacy-risk posed by several types of attacks exists. Different methods such as k-anonymity [18], l-diversity [15], t-closeness [13], and differential privacy [4] have been developed to address the privacy issue. Differential privacy is a formal framework to quantify the degree to which the privacy for each individual in the dataset is preserved while releasing the output of a data analysis algorithm. Differential privacy guarantees that an adversary, by virtue of presence or absence of an individual's data in the dataset, can't draw any conclusions about an individual from the released output of the analysis algorithm. Differential privacy, however, doesn't always adequately limit inference about participation of a single record in the database [6]. Differential privacy requirement does not necessarily constrain the information leakage from a data set [2]. Correlation among records of a dataset would degrade the expected privacy guarantees of differential privacy mechanism [14]. These limitations of differential privacy motivate an information theoretic approach to privacy where privacy is quantified by the mutual information between sensitive information and the released data [1,2,16,17,20].

A data release mechanism aims to provide useful data available while simultaneously limiting any reveled sensitive information. The data perturbation approach uses a random noise adding mechanism to preserve privacy, however, results in distortion of useful data and thus utility of any subsequent machine learning and data analytics algorithm is adversely affected. There remains the challenge of studying and optimizing privacy-utility tradeoff especially in the case when statistical distributions of data are unknown. Information theoretic privacy can be optimized theoretically using a prior knowledge about data statistics. However, in practice, a prior knowledge (such as joint distributions of public and private variables) is missing and therefore a data-driven approach based on generative adversarial networks has been suggested [5]. The data-driven approach of [5] leverages recent advancements in generative adversarial networks to allow learning the parameters of the privatization mechanism. However, the framework of [5] is limited to only binary type of sensitive variables. A similar approach [19] applicable to arbitrary distributions (discrete, continuous, and/or multivariate) of variables employs adversarial training to perform a variational approximation of mutual information privacy. The approach of approximating mutual information via a variational lower bound was also used in [3].

We introduce a novel information theoretic approach for studying privacy-utility tradeoff suitable for multivariate data and for the cases with unknown statistical distributions. The approach is to consider entropy of the noise as a design parameter for studying privacy of a data release mechanism. The privacy-utility tradeoff optimization problem is mathematically formulated where a sample of sensitive or private data x ($x \in \mathcal{X} \subseteq \mathbb{R}^n$), corresponding observed data vector y ($y \in \mathcal{Y} \subseteq \mathbb{R}^p$), and the released data vector z ($z \in \mathcal{Z} \subseteq \mathbb{R}^p$) are modeled as random variables. A privacy-preserving mechanism to release data vector z will add random noise $v \in \mathbb{R}^p$ (sampled from a density function, say $q(v)$) to the observed data vector y, i.e.,

$$z(v; y) = y + v. \tag{1}$$

A relevant optimization problem here is to minimize the privacy-leakage quantified by the mutual information $I(x; z)$ while simultaneously minimizing the amount of data distortion quantified by $E_{q(v)}[\|z(v; y) - y\|^2]$ where $E_{q(v)}[\cdot]$ denotes expectation w.r.t. probability density function $q(v)$. The optimization of tradeoff between minimizing privacy-leakage $I(x; z)$ and minimizing data distortion $E_{q(v)}[\|z(v; y) - y\|^2]$ can be analytically solved for a known data distribution $P_{X,Y}(x, y)$ over the space $\mathcal{X} \times \mathcal{Y}$. The framework proposed in this study, without knowing data distribution, plots the optimized tradeoff curve between $I(x; z)$ and $E_{q(v)}[\|z(v; y) - y\|^2]$. This is done as follows:

1. The probability density function of noise (i.e. $q(v)$), that for a given noise entropy level h, minimizes $E_{q(v)}[\|z(v; y) - y\|^2]$ is analytically derived. In our previous work [12], we have solved a similar optimization problem where the scalar noise distribution minimizing expected magnitude value was derived.
2. The privacy of sensitive data is preserved via adding random noise (sampled from derived optimal distribution) to the data observations, i.e., Eq. (1). Only the noise added data observations are meant to be publicly released.
3. Given a finite set of private-public data pairs $\{(x^i, z^i) \mid i \in \{1, \cdots, N\}\}$, a stochastic fuzzy model $\mathcal{G}$ is built using variational Bayesian methodology such that $x^i = \mathcal{G}(z^i) + v^i$, where $v^i \in \mathbb{R}^n$ is the disturbance vector affecting the data model.
4. A lower bound on privacy-leakage is derived as a functional of distributions characterizing the data model, i.e., $I(x; z) \geq I_L(q(\alpha, \beta))$, where $q(\alpha, \beta)$ is an arbitrary probability density function on parameters α and β which characterizes the distributions related to data model: $x = \mathcal{G}(z) + v$.
5. An approximation to $I(x; z)$ is provided via maximizing I_L w.r.t. $q(\alpha, \beta)$, i.e., $\hat{I}(x; z) = \max_{q(\alpha, \beta)} I_L(q(\alpha, \beta))$.
6. Finally, a curve between $E_{q(v)}[\|z(v; y) - y\|^2]$ and $\hat{I}(x; z)$ is plotted via varying the noise entropy level h.

The aforementioned approach to study and optimize privacy-utility tradeoff is novel. The are three significant features of the proposed framework. First is its generality for any unknown data distribution $P_{X,Y}$, the second is deriving optimal noise adding mechanism analytically, and the third is to compute privacy-leakage analytically without relying on the training of black-box models (e.g. adversarial networks [19]) for approximating distributions. The paper is organized into sections. The optimal noise adding mechanism is derived in Sect. 2 followed by variational Bayesian fuzzy data modeling in Sect. 3. An expression for quantifying privacy-leakage is derived in Sect. 4. A simulation study is provided in Sect. 5 to verify the accuracy of the proposed methodology in optimizing informational privacy. Finally, the concluding remarks are provided in Sect. 6.

2   Deriving Optimal Noise Adding Mechanism

We derive the probability density function of noise (to be added to data observation), that for a given entropy level, minimizes the data distortion function

$E_{q(v)}[\|z(v;y) - y\|^2]$. It follows from (1) that minimization of data distortion function is equivalent to minimizing $E_{q(v)}[\|v\|^2]$ i.e. minimizing the expected squared l_2−norm of the noise vector.

Result 1 (Minimum Squared l_2−norm for a Given Entropy Level). *The probability density function of noise vector that, for a given level of entropy, minimizes the expected squared l_2−norm of noise vector is given as*

$$q^*(v;h) = \left(1/\sqrt{\exp(2h-p)}\right) \exp\left(-\pi \exp\left(1 - (2/p)h\right) \|v\|^2\right), \qquad (2)$$

where h is the given entropy level. The expected squared l_2−norm of noise vector is given as

$$E_{q^*(v;h)}\left[\|v\|^2\right] = (p/(2\pi)) \exp\left((2/p)h - 1\right). \qquad (3)$$

Proof. We seek to solve

$$q^*(v;h) = \arg \min_{q(v)} \int_{\mathbb{R}^p} dv\, q(v) \left(\|v\|^2/2\right), \text{ subject to}$$

$$\int_{\mathbb{R}^p} dv\, q(v) = 1 \qquad (4)$$

$$-\int_{\mathbb{R}^p} dv\, q(v) \log\left(q(v)\right) = h. \qquad (5)$$

Introducing Lagrange multiplier λ_1 for (4) and λ_2 for (5), the following Lagrangian is obtained:

$$\mathcal{L}(q(v), \lambda_1, \lambda_2) = \int_{\mathbb{R}^p} dv\, q(v) \left(\|v\|^2/2\right) + \lambda_1 \left(\int_{\mathbb{R}^p} dv\, q(v) - 1\right)$$
$$+ \lambda_2 \left(h + \int_{\mathbb{R}^p} dv\, q(v) \log\left(q(v)\right)\right).$$

The functional derivative of $\mathcal{L}$ with respect to $q(v)$ is given as

$$\delta\mathcal{L}/\delta q(v) = \left(\|v\|^2/2\right) + \lambda_1 + \lambda_2 \left(1 + \log\left(q(v)\right)\right). \qquad (6)$$

Setting $\delta\mathcal{L}/\delta q(v)$ equal to zero, we have

$$q(v) = \exp\left(-1 - (\lambda_1/\lambda_2)\right) \exp\left(-(\|v\|^2)/(2\lambda_2)\right), \ \lambda_2 \neq 0. \qquad (7)$$

Setting $\partial\mathcal{L}/\partial\lambda_1$ equal to zero and then solving using (7), we get

$$q(v) = \left(1/\sqrt{(2\pi)^p(\lambda_2)^p}\right) \exp\left(-(\|v\|^2)/(2\lambda_2)\right), \ \lambda_2 > 0. \qquad (8)$$

Setting $\partial \mathcal{L}/\partial \lambda_2$ equal to zero and then solving using (8), we get the optimal value of λ_2 as

$$\lambda_2^* = (1/(2\pi)) \exp\left((2/p)h - 1\right). \tag{9}$$

Using the optimal value of λ_2^* in (8), the optimal expression for $q(v)$ is obtained as in (2). As $\lambda_2^* > 0$, $\mathcal{L}$ is convex in $q(v)$ and thus $q^*(v)$ corresponds to the minimum. Finally, the expected value of squared l_2-norm of v with respect to $q^*(v)$ is given by (3).

3   Variational Bayesian Fuzzy Data Modeling

Consider a Takagi-Sugeno fuzzy filter $(\mathcal{F} : \mathbb{R}^q \rightarrow \mathbb{R})$ that maps $q-$dimensional real-space to $1-$dimensional real-line. The fuzzy filter consists of M number of rules of following type:

$$\text{If } s \text{ is } \mathbf{A}_m, \text{ then } \mathcal{F}(s) = c_m, \ m \in \{1, \cdots, M\}$$

where $s \in \mathbb{R}^q$, $c_m \in \mathbb{R}$, and the fuzzy set $\mathbf{A}_m$ is defined, without loss of generality, with the following Gaussian membership function

$$\mu_{\mathbf{A}_m}(s) = \exp\left(-0.5 \|s - a^m\|_W^2\right) \tag{10}$$

where $a^m \in \mathbb{R}^q$ is the mean of $\mathbf{A}_m$, $W \in \mathbb{R}^{q \times q}(W > 0)$, and $\|s\|_P^2 \overset{\text{def}}{=} s^T P s$. For a given input $s \in \mathbb{R}^q$, the *degree of fulfillment* of the $m-$th rule is given by $\mu_{\mathbf{A}_m}(s)$. The output of the filter to input vector s is computed by taking the weighted average of the output provided by each rule, i.e.,

$$\mathcal{F}(s) = \frac{\sum_{m=1}^{M} \mu_{\mathbf{A}_m}(s) c_m}{\sum_{m=1}^{M} \mu_{\mathbf{A}_m}(s)}. \tag{11}$$

Definition 1 (A Stochastic Fuzzy Model (FM)). *A stochastic fuzzy model, $\mathcal{G} : \mathbb{R}^p \rightarrow \mathbb{R}^n$, maps an input vector $z \in \mathbb{R}^p$ to the output vector $\mathcal{G}(z) \in \mathbb{R}^n$ given as*

$$\mathcal{G}(z) = \left[\mathcal{F}_1(V^T z) \cdots \mathcal{F}_n(V^T z)\right]^T \in \mathbb{R}^n \tag{12}$$

where $V \in \mathbb{R}^{p \times q}$ (with $q \leq p$) is a matrix, $\mathcal{F}_k$ ($k \in \{1, 2, \cdots, n\}$) is a Takagi-Sugeno fuzzy filter (11), with consequent parameters being considered as random variables and being represented by $\alpha_k = \left[c_{k,1} \cdots c_{k,M}\right]^T \in \mathbb{R}^M$, such that

$$\mathcal{F}_k(s) = \frac{\sum_{m=1}^{M} \mu_{\mathbf{A}_m}(s) c_{k,m}}{\sum_{m=1}^{M} \mu_{\mathbf{A}_m}(s)}. \tag{13}$$

Given a finite set of input-output pairs $\mathcal{D} = \{(x^i, z^i) \mid i \in \{1, \cdots, N\}\}$, the data is modeled through a stochastic fuzzy model $\mathcal{G}$ as

$$x^i = \mathcal{G}(z^i) + v^i \tag{14}$$

$$= \left[\mathcal{F}_1(V^T z^i) \cdots \mathcal{F}_n(V^T z^i) \right]^T + v^i. \tag{15}$$

The following notation is introduced:

$$\mathbf{z} = \left\{ z^i \mid z^i \in \mathbb{R}^p, i \in \{1, \cdots, N\} \right\} \tag{16}$$

$$\mathbf{a} = \left\{ a^m \mid a^m \in \mathbb{R}^q, m \in \{1, \cdots, M\} \right\} \tag{17}$$

$$\mathbf{f}_k = \left[\mathcal{F}_k(V^T z^1) \cdots \mathcal{F}_k(V^T z^N) \right]^T \in \mathbb{R}^N \tag{18}$$

$$\alpha_k = \left[c_{k,1} \cdots c_{k,M} \right]^T \in \mathbb{R}^M \tag{19}$$

$$\mathbf{x}_k = \left[x_k^1 \cdots x_k^N \right]^T \in \mathbb{R}^N \tag{20}$$

$$\mathbf{v}_k = \left[v_k^1 \cdots v_k^N \right]^T \in \mathbb{R}^N \tag{21}$$

where $k \in \{1, \cdots, n\}$, and x_k^i and v_k^i denote the $k-$th element of x^i and v^i respectively. Let $K_{\mathbf{za}} \in \mathbb{R}^{N \times M}$ be a matrix whose $(i, m)-$th element is given as

$$(K_{\mathbf{za}}(V, W))_{i,m} = \frac{\exp\left(-0.5 \left\| V^T z^i - a^m \right\|_W^2 \right)}{\sum_{m=1}^M \exp\left(-0.5 \left\| z^i - a^m \right\|_W^2 \right)}. \tag{22}$$

It follows from (13), (22), and (19) that

$$\mathbf{f}_k = K_{\mathbf{za}} \alpha_k. \tag{23}$$

Also, it can be observed that

$$\mathbf{x}_k = K_{\mathbf{za}} \alpha_k + \mathbf{v}_k. \tag{24}$$

The disturbance vector $\mathbf{v}_k$ is priori assumed to be Gaussian with mean zero and a precision of β, i.e.,

$$p(\mathbf{v}_k | \beta) = \left(1/\sqrt{(2\pi)^N (\beta)^{-N}} \right) \exp\left(-0.5\beta \|\mathbf{v}_k\|^2 \right) \tag{25}$$

where $\beta > 0$ is priori assumed to be Gamma distributed:

$$p(\beta; a, b) = (b^a/\Gamma(a))\,(\beta)^{a-1}\exp(-b\beta) \tag{26}$$

where $a, b > 0$. The Gaussian prior is taken over parameter vector α_k:

$$p(\alpha_k; m_k, \Lambda_k) = \left(1/\sqrt{(2\pi)^M |(\Lambda_k)^{-1}|}\right)\exp\left(-0.5(\alpha_k - m_k)^T \Lambda_k(\alpha_k - m_k)\right) \tag{27}$$

where $m_k \in \mathbb{R}^M$ and $\Lambda_k \in \mathbb{R}^{M \times M}(\Lambda_k > 0)$. Define sets

$$\mathbf{X} \stackrel{\text{def}}{=} \{x_1, \cdots, x_n\} \tag{28}$$

$$\alpha \stackrel{\text{def}}{=} \{\alpha_1, \cdots, \alpha_n\} \tag{29}$$

and consider the marginal probability of data $\mathbf{X}$ which is given as

$$p(\mathbf{X}) = \int d\alpha\, d\beta\, p(\mathbf{X}, \alpha, \beta). \tag{30}$$

Let $q(\alpha, \beta)$ be an arbitrary distribution. The log marginal probability of $\mathbf{X}$ can be expressed as

$$
\log(p(\mathbf{X})) = \int d\alpha\, d\beta\, q(\alpha, \beta) \log\left(\frac{p(\mathbf{X}, \alpha, \beta)}{q(\alpha, \beta)}\right)
$$
$$
+ \int d\alpha\, d\beta\, q(\alpha, \beta) \log\left(\frac{q(\alpha, \beta)}{p(\alpha, \beta|\mathbf{X})}\right). \tag{31}
$$

Define

$$F(q(\alpha, \beta), \mathbf{X}) \stackrel{\text{def}}{=} \int d\alpha\, d\beta\, q(\alpha, \beta) \log\left(p(\mathbf{X}, \alpha, \beta)/q(\alpha, \beta)\right) \tag{32}$$

to express (31) as

$$\log(p(\mathbf{X})) = F(q(\alpha, \beta), \mathbf{X}) + \text{KL}(q(\alpha, \beta)\|p(\alpha, \beta|\mathbf{X})) \tag{33}$$

where KL is the Kullback-Leibler divergence of $p(\alpha, \beta|\mathbf{X})$ from $q(\alpha, \beta)$ and F, referred to as negative free energy, provides a lower bound on the logarithmic evidence for the data.

The variational Bayesian approach minimizes the difference (in term of KL divergence) between variational and true posteriors via analytically maximizing negative free energy F over variational distributions. However, the analytical derivation requires the following widely used mean-field approximation:

$$q(\alpha, \beta) = q(\alpha)q(\beta) \tag{34}$$

$$= q(\alpha_1)\cdots q(\alpha_n)q(\beta). \tag{35}$$

Applying the standard variational optimization technique (as in [7–11]), it can be verified that the optimal variational distributions maximizing F are as follows:

$$q^*(\alpha_k) = \left(1/\sqrt{(2\pi)^M|(\hat{\Lambda}_k)^{-1}|}\right) \exp\left(-0.5(\alpha_k - \hat{m}_k)^T \hat{\Lambda}_k(\alpha_k - \hat{m}_k)\right) \quad (36)$$

$$q^*(\beta) = \left((\hat{b})^{\hat{a}}/\Gamma(\hat{a})\right)(\beta)^{\hat{a}-1}\exp(-\hat{b}\beta) \quad (37)$$

where the parameters $(\hat{\Lambda}_k, \hat{m}_k, \hat{a}, \hat{b})$ satisfy the following:

$$\hat{\Lambda}_k = \Lambda_k + \left(\hat{a}/\hat{b}\right)(K_{za})^T K_{za} \quad (38)$$

$$\hat{m}_k = (\hat{\Lambda}_k)^{-1}\left(\Lambda_k m_k + \left(\hat{a}/\hat{b}\right)(K_{za})^T x_k\right) \quad (39)$$

$$\hat{a} = a + 0.5nN \quad (40)$$

$$\hat{b} = b + 0.5\sum_{k=1}^{n}\left\{\|x_k - K_{za}\hat{m}_k\|^2 + Tr\left((\hat{\Lambda}_k)^{-1}(K_{za})^T K_{za}\right)\right\} \quad (41)$$

where $\mathrm{Tr}(\cdot)$ denotes the trace operator.

Algorithm 1. An algorithm for variational Bayesian inference of data model.

Require: Data set $\mathcal{D} = \{(x^i, z^i) \mid x^i \in \mathbb{R}^n, z^i \in \mathbb{R}^p, i \in \{1, \cdots, N\}\}$, the number of rules in a fuzzy filter $M \in \mathbb{Z}_+$, the subspace dimension $q \in \mathbb{Z}_+$ with $q \le p$.

1: Define $V \in \mathbb{R}^{p \times q}$ such that j−th column of V is equal to eigenvector corresponding to j−th largest eigenvalue of sample covariance matrix of $\{z^i \mid i \in \{1, \cdots, N\}\}$.

2: Compute $s^i = V^T z^i$, for $i \in \{1, \cdots, N\}$.

3: The fuzzy sets' mean values, $\mathbf{a} = \{a^m \mid m \in \{1, \cdots, M\}\}$, are defined as

$$\{a^m \mid m \in \{1, \cdots, M\}\} = \mathrm{ClusterCentroid}(\{s^i \mid i \in \{1, \cdots, N\}\}, M) \quad (42)$$

ClusterCentroid$(\cdot)$ represents the k-means clustering to return M cluster centroids.

4: Define W to be a diagonal matrix such that j−th diagonal element is equal to the inverse of squared-distance between two most-distant points in the set $\{s^i_j \mid i \in \{1, \cdots, N\}\}$, where s^i_j is j−th element of s^i.

5: Compute $K_{za}(V, W)$ using (22).

6: Choose $a = 10^{-6}, b = 10^{-6}, m_j = 0, \Lambda_j = 10^{-6}I$.

7: Initialise $\hat{a}/\hat{b} = 1$.

8: **repeat**

9:    Update $\{\hat{\Lambda}_k, \hat{m}_k \mid k \in \{1, \cdots, n\}\}, \hat{a}, \hat{b}$ using (38), (39), (40), (41).

10: **until** (convergence **or** iterations = 1000)

11: **return** $\mathcal{M} = \{\hat{a}, \hat{b}, \{\hat{m}_k, \hat{\Lambda}_k \mid k \in \{1, \cdots, n\}\}, \mathbf{a}, V, W\}$.

Variational Bayesian inference lends itself to a data modeling algorithm formally stated as Algorithm 1. The optimal distributions $q^*(\alpha_k)$ and $q^*(\beta)$ determined using Algorithm 1 define a model as stated in Remark 1.

Remark 1 (Model). The model built using Algorithm 1 relates sensitive data vector $x = [\, x_1 \cdots x_n \,]^T \in \mathbb{R}^n$ to released data vector $z \in \mathbb{R}^p$ as

$$x_k = k(z)\alpha_k + v_k, \tag{43}$$

$$p(v_k|\beta) = \left(1/\sqrt{(2\pi)(\beta)^{-1}}\right) \exp\left(-0.5\beta|v_k|^2\right), \tag{44}$$

$$p(\beta; \hat{a}, \hat{b}) = \left(\hat{b}^{\hat{a}}/\Gamma(\hat{a})\right)(\beta)^{\hat{a}-1} \exp(-\hat{b}\beta), \tag{45}$$

$$p(\alpha_k; \hat{m}_k, \hat{\Lambda}_k) = \left(1/\sqrt{(2\pi)^M |(\Lambda_k)^{-1}|}\right) \exp\left(-0.5(\alpha_k - \hat{m}_k)^T \hat{\Lambda}_k (\alpha_k - \hat{m}_k)\right) \tag{46}$$

where $k(z) \in \mathbb{R}^{1 \times M}$ is a vector-valued function whose $m-$th element is given as

$$(k(z))_m = \frac{\exp\left(-0.5 \left\|V^T z - a^m\right\|_W^2\right)}{\sum_{m=1}^M \exp\left(-0.5 \left\|V^T z - a^m\right\|_W^2\right)}. \tag{47}$$

Here, $\{\hat{a}, \hat{b}, \{\hat{m}_k, \hat{\Lambda}_k \mid k \in \{1, \cdots, n\}\}, \mathbf{a}, V, W\}$ are returned by Algorithm 1.

4   Variational Approximation of Informational Privacy

The mutual information between sensitive data vector x and released data vector z is given as

$$I(x; z) = H(x) - H(x|z) \tag{48}$$

$$= H(x) + \int_{X, Z} p(x, z) \log\left(p(x|z)\right) \mathrm{d}x\, \mathrm{d}z \tag{49}$$

$$= H(x) + \langle\log\left(p(x|z)\right)\rangle_{p(x,z)} \tag{50}$$

where $H(x)$, $H(x|z)$ are marginal, conditional entropies respectively and the averaging operator $< \cdot >.$ is defined as

$$<f(x)>_{p(x)} = \int \mathrm{d}x\, p(x)f(x). \tag{51}$$

Result 2 (Variational Approximation of $I(x; z)$). *Assuming the data model as stated in Remark 1, a variational approximation of $I(x; z)$ is given as*

$$\hat{I}(x; z) = \tag{52}$$
$$H(x) - 0.5n \log(2\pi) + 0.5n \left\{ \Psi(\bar{a}) - \log(\bar{b}) \right\}$$
$$- \frac{1}{2} \left(\bar{a}/\bar{b}\right) \sum_{k=1}^{n} \left\langle |x_k - k(z)\bar{\mathrm{m}}_k|^2 \right\rangle_{p(x,z)} - \frac{1}{2} \left(\bar{a}/\bar{b}\right) \sum_{k=1}^{n} \left\langle Tr \left((\bar{\Lambda}_k)^{-1}(k(z))^T k(z)\right) \right\rangle_{p(z)}$$
$$- 0.5 \sum_{k=1}^{n} \left\{ (\hat{\mathrm{m}}_k - \bar{\mathrm{m}}_k)^T \hat{\Lambda}_k (\hat{\mathrm{m}}_k - \bar{\mathrm{m}}_k) + Tr \left(\hat{\Lambda}_k (\bar{\Lambda}_k)^{-1} \right) - \log \left(\frac{|(\bar{\Lambda}_k)^{-1}|}{|(\hat{\Lambda}_k)^{-1}|} \right) \right\}$$
$$+ 0.5nM - \hat{a} \log \left(\bar{b}/\hat{b} \right) + \log \left(\Gamma(\bar{a})/\Gamma(\hat{a}) \right) - (\bar{a} - \hat{a})\Psi(\bar{a}) + (\bar{b} - \hat{b}) \left(\bar{a}/\bar{b} \right)$$

where $\Psi(\cdot)$ is the digamma function and the parameters $(\bar{\Lambda}_k, \bar{\mathrm{m}}_k, \bar{a}, \bar{b})$ satisfy followings:

$$\bar{\Lambda}_k = \hat{\Lambda}_k + \left(\bar{a}/\bar{b} \right) \left\langle (k(z))^T k(z) \right\rangle_{p(z)} \tag{53}$$

$$\bar{\mathrm{m}}_k = (\bar{\Lambda}_k)^{-1} \left(\hat{\Lambda}_k \hat{\mathrm{m}}_k + \left(\bar{a}/\bar{b} \right) \left\langle (k(z))^T x_k \right\rangle_{p(x,z)} \right) \tag{54}$$

$$\bar{a} = \hat{a} + 0.5n \tag{55}$$

$$\bar{b} = \hat{b} + \frac{1}{2} \sum_{k=1}^{n} \left\langle |x_k - k(z)\bar{\mathrm{m}}_k|^2 \right\rangle_{p(x,z)}$$
$$+ \frac{1}{2} \sum_{k=1}^{n} \left\langle Tr \left((\bar{\Lambda}_k)^{-1}(k(z))^T k(z) \right) \right\rangle_{p(z)} \tag{56}$$

Proof. Consider the conditional probability of x which is given as

$$p(x|z) = \int d\alpha \, d\beta \, p(\alpha, \beta, x|z) \tag{57}$$

where α is a set defined as in (29). Let $q(\alpha, \beta)$ be an arbitrary distribution. The log conditional probability of x can be expressed as

$$\log(p(x|z)) = \int d\alpha \, d\beta \, q(\alpha, \beta) \log \left(p(x|z) \right) \tag{58}$$

$$= \int d\alpha \, d\beta \, q(\alpha, \beta) \log \left(p(\alpha, \beta, x|z)/p(\alpha, \beta|x, z) \right) \tag{59}$$

$$= \int d\alpha \, d\beta \, q(\alpha, \beta) \log \left(p(\alpha, \beta, x|z)/q(\alpha, \beta) \right)$$
$$+ \int d\alpha \, d\beta \, q(\alpha, \beta) \log \left(q(\alpha, \beta)/p(\alpha, \beta|x, z) \right). \tag{60}$$

Define

$$F(q(\alpha, \beta), x, z) \overset{\text{def}}{=} \int d\alpha\, d\beta\, q(\alpha, \beta) \log\left(p(\alpha, \beta, x|z)/q(\alpha, \beta)\right) \qquad (61)$$

to express (60) as

$$\log(p(x|z)) = F(q(\alpha, \beta), x, z) + \mathrm{KL}(q(\alpha, \beta)\|p(\alpha, \beta|x, z)) \qquad (62)$$

where KL is Kullback-Leibler divergence of $p(\alpha, \beta|x, z)$ from $q(\alpha, \beta)$. Using (50),

$$I(x; z) = H(x) + \langle F(q(\alpha, \beta), x, z)\rangle_{p(x,z)} + \langle \mathrm{KL}(q(\alpha, \beta)\|p(\alpha, \beta|x, z))\rangle_{p(x,z)}. \qquad (63)$$

Since Kullback–Leibler divergence is always non-zero, it follows from (63) that $H(x) + \langle F\rangle_{p(x,z)}$ provides a lower bound on $I(x; z)$ i.e.

$$I(x; z) \geq H(x) + \langle F(q(\alpha, \beta), x, z)\rangle_{p(x,z)}. \qquad (64)$$

Our approach to approximate $I(x; z)$ is to maximize its lower bound with respect to variational distribution $q(\alpha, \beta)$. That is, we solve

$$\hat{I}(x; z) = \max_{q(\alpha,\beta)} \left(H(x) + \langle F(q(\alpha, \beta), x, z)\rangle_{p(x,z)} \right) \qquad (65)$$

$$= H(x) + \max_{q(\alpha,\beta)} \langle F(q(\alpha, \beta), x, z)\rangle_{p(x,z)}. \qquad (66)$$

For this, consider

$$F(q(\alpha, \beta), x, z) = \langle \log(p(x|\alpha, \beta, z))\rangle_{q(\alpha,\beta)} + \langle \log\left(p(\alpha, \beta)/q(\alpha, \beta)\right)\rangle_{q(\alpha,\beta)}. \qquad (67)$$

Assuming that $x_1, \cdots, x_n$ are independent,

$$\log(p(x|z, \alpha, \beta)) = \sum_{k=1}^{n} \log(p(x_k|z, \alpha_k, \beta)). \qquad (68)$$

It follows from (44) and (43) that

$$\log(p(x_k|z, \alpha_k, \beta)) = -0.5\log(2\pi) + 0.5\log(\beta) - 0.5\beta|x_k - k(z)\alpha_k|^2. \qquad (69)$$

Using (68), (69), and (34–35) in (67), we have

$$F = -0.5n\log(2\pi) + 0.5n\,\langle\log(\beta)\rangle_{q(\beta)} - 0.5\,\langle\beta\rangle_{q(\beta)}\sum_{k=1}^{n}\langle|x_k - k(z)\alpha_k|^2\rangle_{q(\alpha_k)}$$

$$+ \sum_{k=1}^{n}\left\langle\log\left(p(\alpha_k; \hat{m}_k, \hat{\Lambda}_k)/q(\alpha_k)\right)\right\rangle_{q(\alpha_k)} + \left\langle\log\left(p(\beta; \hat{a}, \hat{b})/q(\beta)\right)\right\rangle_{q(\beta)}. \qquad (70)$$

Thus,

$$
\langle F \rangle_{p(x,z)} = -0.5n \log(2\pi) + 0.5n \langle \log(\beta) \rangle_{q(\beta)} - 0.5 \langle \beta \rangle_{q(\beta)} \sum_{k=1}^{n} \langle |x_k|^2 \rangle_{p(x)}
$$

$$
- 0.5 \langle \beta \rangle_{q(\beta)} \sum_{k=1}^{n} \left\langle (\alpha_k)^T \langle (k(z))^T k(z) \rangle_{p(z)} \alpha_k \right\rangle_{q(\alpha_k)}
$$

$$
+ \langle \beta \rangle_{q(\beta)} \sum_{k=1}^{n} \left\langle (\alpha_k)^T \langle (k(z))^T x_k \rangle_{p(x,z)} \right\rangle_{q(\alpha_k)}
$$

$$
+ \sum_{k=1}^{n} \left\langle \log \left(p(\alpha_k; \hat{\mathrm{m}}_k, \hat{\Lambda}_k)/q(\alpha_k) \right) \right\rangle_{q(\alpha_k)}
$$

$$
+ \left\langle \log \left(p(\beta; \hat{a}, \hat{b})/q(\beta) \right) \right\rangle_{q(\beta)}. \tag{71}
$$

Now, $\langle F \rangle_{p(x,z)}$ can be maximized w.r.t. $q(\alpha_k)$ and $q(\beta)$ using variational optimization. It can be seen that optimal distributions maximizing $\langle F \rangle_{p(x,z)}$ are given as

$$
q^*(\alpha_k) = \left(1/\sqrt{(2\pi)^M |(\bar{\Lambda}_k)^{-1}|} \right) \exp \left(-0.5(\alpha_k - \bar{\mathrm{m}}_k)^T \bar{\Lambda}_k (\alpha_k - \bar{\mathrm{m}}_k) \right) \tag{72}
$$

$$
q^*(\beta) = \left((\bar{b})^{\bar{a}}/\Gamma(\bar{a}) \right) (\beta)^{\bar{a}-1} \exp(-\bar{b}\beta) \tag{73}
$$

where the parameters $(\bar{\Lambda}_j^l, \bar{\mathrm{m}}_j^l, \bar{a}_l, \bar{b}_l)$ satisfy (53), (54), (55), (56). The maximum attained value of $\langle F \rangle_{p(x,z)}$ is given as

$$
\langle F(q^*(\alpha, \beta), x, z) \rangle_{p(x,z)} = \tag{74}
$$

$$
-0.5n \log(2\pi) + 0.5n \left\{ \Psi(\bar{a}) - \log(\bar{b}) \right\} - 0.5 \left(\bar{a}/\bar{b} \right) \sum_{k=1}^{n} \langle |x_k - k(z)\bar{\mathrm{m}}_k|^2 \rangle_{p(x,z)}
$$

$$
-0.5 \left(\bar{a}/\bar{b} \right) \sum_{k=1}^{n} \langle Tr \left((\bar{\Lambda}_k)^{-1} (k(z))^T k(z) \right) \rangle_{p(z)} - \sum_{k=1}^{n} \mathrm{KL}(q^*(\alpha_k) \| p(\alpha_k; \hat{\mathrm{m}}_k, \hat{\Lambda}_k))
$$

$$
-\mathrm{KL}(q^*(\beta) \| p(\beta; \hat{a}, \hat{b}))
$$

where $\Psi(\cdot)$ is the digamma function. After substituting the maximum value of $\langle F \rangle_{p(x,z)}$ in (66) and calculating Kullback-Leibler divergences, we get (52).

Result 2 can be leveraged to study and optimize privacy-utility tradeoff. Algorithm 2 outlines the proposed information theoretic approach to privacy-utility tradeoff optimization.

5   A Simulation Study

To demonstrate the proposed framework, a scenario is considered where sensitive data $x \in \mathbb{R}^{10}$ is Gaussian distributed such that $x \sim \mathcal{N}(0, 5I_{10})$, and the observed

Algorithm 2. An algorithm for plotting optimized tradeoff curve between informational privacy and data distortion.

Require: Data set $\{(x^i, y^i) \mid x^i \in \mathbb{R}^n, y^i \in \mathbb{R}^p, i \in \{1, \cdots, N\}\}$.

1: Consider a finite set of equidistanced points over a range of noise entropy level, i.e., $\mathbf{h} = \{h_j \mid j \in \{1, \cdots, L\}, h_{min} < h_1 < \cdots < h_L < h_{max}\}$.

2: **for** each $h_j \in \mathbf{h}$ **do**

3:     Choose N random samples of noise vectors such that i−th noise sample $v^i \sim q^*(v; h_j)$, where $q^*(v; h_j)$ is defined as in (2).

4:     Add noise to observed data, i.e. $z^i = y^i + v^i$.

5:     Apply Algorithm 1 on $\{(x^i, z^i) \mid i \in \{1, \cdots, N\}\}$ taking e.g. $M = \lceil N/2 \rceil$ and $q = \min(20, p)$ to build the data model $\mathcal{M} = \{\hat{a}, \hat{b}, \{\hat{m}_k, \hat{\Lambda}_k \mid k \in \{1, \cdots, n\}\}, \mathbf{a}, V, W\}$.

6:     Initialise $\bar{a}/\bar{b} = \hat{a}/\hat{b}$.

7:     **repeat**

8:         Update $\{\bar{\Lambda}_k, \bar{m}_k \mid k \in \{1, \cdots, n\}\}, \bar{a}, \bar{b}$ using (53), (54), (55), (56).

9:     **until** (convergence **or** iterations $= 1000$)

10:     Compute $pl_j = \hat{I}(x; z) - H(x)$ using (52). As entropy of private data $H(x)$ remains fixed independent of noise adding mechanism, pl_j is a measure of $\hat{I}(x; z)$ and thus pl_j quantifies privacy-leakage from sensitive data x to released public data z.

11:     Compute data distortion value $dd_j = E_{q^*(v; h_j)} [\|v\|^2]$ using (3).

12: **end for**

13: Draw privacy-leakage vs. data-distortion curve via plotting the points in the set $\{(dd_j, pl_j) \mid j \in \{1, \cdots, L\}\}$.

---

data y is same as the sensitive data i.e. $y = x$. The dataset consists of $N(= 1000)$ samples of private and observed data such that i−th private data sample $x^i \sim \mathcal{N}(0, 5I_{10})$ and i−th observed data sample $y^i = x^i$. Algorithm 2 is applied taking 25 uniformly distributed points over $[-20, 5]$ as noise entropy levels to plot optimized tradeoff curve between privacy-leakage and data-distortion.

Since the data distribution in this scenario is known, the optimal privacy-utility tradeoff curve can be theoretically determined and is given as $I(x; z) = \sum_{k=1}^{10} \max(0, 0.5 \log(5/\delta))$, where δ is the distortion in an element of y, i.e., $\delta = E_{q(v)}[\|z_k(v; y) - y_k\|^2]$, $\forall k \in \{1, \cdots, 10\}$. Also the entropy of sensitive data can be theoretically calculated as $H(x) = 0.5 \log(|(2\pi e 5 I_{10})|)$. Thus, the quantity $I(x; z) - H(x)$ can be theoretically calculated and plotted against data-distortion (which is given as $E_{q(v)}[\|z(v; y) - y\|^2] = 10\delta$). The theoretical privacy-leakage vs. data-distortion curve can be used to evaluate the accuracy of Result 2 and Algorithm 2.

A close agreement between Algorithm 2 and the analytically derived solution verifies the accuracy of the proposed method in optimizing informational privacy.

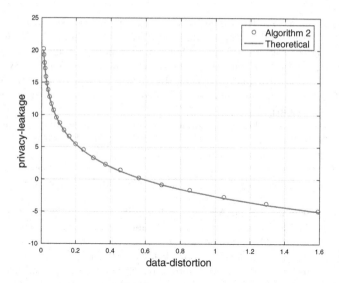

Fig. 1. A comparison of the optimized tradeoff curve between privacy-leakage ($\hat{I}(x; z) - H(x)$) and data-distortion ($E_{q^*(v)}\left[\|z(v; y) - y\|^2\right]$) obtained via Algorithm 2 with the analytically derived curve using data distribution knowledge.

6   Concluding Remarks

This study has outlined a novel information theoretic approach to variational approximation of informational privacy. The introduced framework facilitates to study and optimize the privacy-utility tradeoff where privacy-leakage is quantified by the mutual information between sensitive data and released data. A simulation study has verified that the privacy-utility tradeoff curve computed via proposed method matches closely to the analytically derived curve for a known data distribution (Fig. 1).

This paper has considered expected squared l_2−norm as the data-distortion function, however, the framework is general allowing any other data-distortion function such as expected l_1−norm or Kullback-Leibler divergence.

References

1. Basciftci, Y.O., Wang, Y., Ishwar, P.: On privacy-utility tradeoffs for constrained data release mechanisms. In: 2016 Information Theory and Applications Workshop (ITA), pp. 1–6, January 2016. https://doi.org/10.1109/ITA.2016.7888175
2. Calmon, F.D.P., Fawaz, N.: Privacy against statistical inference. In: Proceedings of the 50th Annual Allerton Conference on Communication, Control, and Computing, Allerton (2012). http://arxiv.org/abs/1210.2123

3. Chen, X., Duan, Y., Houthooft, R., Schulman, J., Sutskever, I., Abbeel, P.: InfoGAN: Interpretable representation learning by information maximizing generative adversarial nets. In: Lee, D.D., Sugiyama, M., Luxburg, U.V., Guyon, I., Garnett, R. (eds.) Advances in Neural Information Processing Systems, vol. 29, pp. 2172–2180. Curran Associates, Inc. (2016). http://papers.nips.cc/paper/6399-infogan-interpretable-representation-learning-by-information-maximizing-generative-adversarial-nets.pdf
4. Dwork, C., Roth, A.: The algorithmic foundations of differential privacy. Found. Trends Theor. Comput. Sci. **9**(3–4), 211–407 (2014). https://doi.org/10.1561/0400000042
5. Huang, C., Kairouz, P., Chen, X., Sankar, L., Rajagopal, R.: Context-aware generative adversarial privacy. Entropy **19**(12), 656 (2017). https://doi.org/10.3390/e19120656
6. Kifer, D., Machanavajjhala, A.: No free lunch in data privacy. In: Proceedings of the 2011 ACM SIGMOD International Conference on Management of Data, pp. 193–204. SIGMOD 2011, Association for Computing Machinery, New York, NY, USA (2011). https://doi.org/10.1145/1989323.1989345
7. Kumar, M., Insan, A., Stoll, N., Thurow, K., Stoll, R.: Stochastic fuzzy modeling for ear imaging based child identification. IEEE Trans. Syst. Man Cybern. Syst. **46**(9), 1265–1278 (2016). https://doi.org/10.1109/TSMC.2015.2468195
8. Kumar, M., et al.: Stress monitoring based on stochastic fuzzy analysis of heartbeat intervals. IEEE Trans. Fuzzy Syst. **20**(4), 746–759 (2012). https://doi.org/10.1109/TFUZZ.2012.2183602
9. Kumar, M., Stoll, N., Stoll, R.: Variational Bayes for a mixed stochastic/deterministic fuzzy filter. IEEE Trans. Fuzzy Syst. **18**(4), 787–801 (2010). https://doi.org/10.1109/TFUZZ.2010.2048331
10. Kumar, M., Stoll, N., Stoll, R.: Stationary fuzzy Fokker-Planck learning and stochastic fuzzy filtering. IEEE Trans. Fuzzy Syst. **19**(5), 873–889 (2011). https://doi.org/10.1109/TFUZZ.2011.2148724
11. Kumar, M., Stoll, N., Stoll, R., Thurow, K.: A stochastic framework for robust fuzzy filtering and analysis of signals-Part I. IEEE Trans. Cybern. **46**(5), 1118–1131 (2016). https://doi.org/10.1109/TCYB.2015.2423657
12. Kumar, M., Rossbory, M., Moser, B.A., Freudenthaler, B.: Deriving an optimal noise adding mechanism for privacy-preserving machine learning. In: Anderst-Kotsis, G., et al. (eds.) Database and Expert Systems Applications, pp. 108–118. Springer International Publishing, Cham (2019)
13. Li, N., Li, T., Venkatasubramanian, S.: t-closeness: privacy beyond k-anonymity and l-diversity. In: Chirkova, R., Dogac, A., Özsu, M.T., Sellis, T.K. (eds.) Proceedings of the 23rd International Conference on Data Engineering, ICDE 2007, The Marmara Hotel, Istanbul, Turkey, 15–20 April 2007, pp. 106–115. IEEE Computer Society (2007). https://doi.org/10.1109/ICDE.2007.367856
14. Liu, C., Chakraborty, S., Mittal, P.: Dependence makes you vulnberable: differential privacy under dependent tuples. In: 23rd Annual Network and Distributed System Security Symposium, NDSS 2016, San Diego, California, USA, 21–24 February 2016. The Internet Society (2016). http://wp.internetsociety.org/ndss/wp-content/uploads/sites/25/2017/09/dependence-makes-you-vulnerable-differential-privacy-under-dependent-tuples.pdf
15. Machanavajjhala, A., Kifer, D., Gehrke, J., Venkitasubramaniam, M.: L-diversity: privacy beyond k-anonymity. ACM Trans. Knowl. Discov. Data **1**(1) (2007). https://doi.org/10.1145/1217299.1217302

16. Rebollo-Monedero, D., Forné, J., Domingo-Ferrer, J.: From t-closeness-like privacy to postrandomization via information theory. IEEE Trans. Knowl. Data Eng. **22**(11), 1623–1636 (2010). https://doi.org/10.1109/TKDE.2009.190

17. Sankar, L., Rajagopalan, S.R., Poor, H.V.: Utility-privacy tradeoffs in databases: an information-theoretic approach. IEEE Trans. Inf. Forensics Secur. **8**(6), 838–852 (2013). https://doi.org/10.1109/TIFS.2013.2253320

18. Sweeney, L.: K-anonymity: a model for protecting privacy. Int. J. Uncertainity Fuzziness Knowl. Based Syst. **10**(5), 557–570 (2002). https://doi.org/10.1142/S0218488502001648

19. Tripathy, A., Wang, Y., Ishwar, P.: Privacy-preserving adversarial networks. In: 2019 57th Annual Allerton Conference on Communication, Control, and Computing (Allerton), pp. 495–505, September 2019. https://doi.org/10.1109/ALLERTON.2019.8919758

20. Wang, Y., Basciftci, Y.O., Ishwar, P.: Privacy-utility tradeoffs under constrained data release mechanisms. CoRR abs/1710.09295 (2017). http://arxiv.org/abs/1710.09295

An Architecture for Automated Security Test Case Generation for MQTT Systems

Hannes Sochor$^{(\boxtimes)}$, Flavio Ferrarotti , and Rudolf Ramler

Software Competence Center Hagenberg,
Softwarepark 21, 4232 Hagenberg, Austria
{hannes.sochor,flavio.ferrarotti,rudolf.ramler}@scch.at

Abstract. Message Queuing Telemetry Transport (MQTT) protocol is among the preferred publish/subscribe protocols used for Machine-to-Machine (M2M) communication and Internet of Things (IoT). Although the MQTT protocol itself is quite simple, the concurrent iteration of brokers and clients and its intrinsic non-determinism, coupled with the diversity of platforms and programming languages in which the protocol is implemented and run, makes the necessary task of security testing challenging. We address precisely this problem by proposing an architecture for security test generation for systems relying on the MQTT protocol. This architecture enables automated test case generation to reveal vulnerabilities and discrepancies between different implementations. As a desired consequence, when implemented, our architectural design can be used to uncover erroneous behaviours that entail latent security risks in MQTT broker and client implementations. In this paper we describe the key components of our architecture, our prototypical implementation using a random test case generator, core design decisions and the use of security attacks in testing. Moreover, we present first evaluations of the architectural design and the prototypical implementation with encouraging initial results.

Keywords: Security testing · Automated testing · IoT · MQTT

1 Introduction

The Internet of Things (IoT) connects millions of devices of different cyber-physical systems (CPSs). However, the increasing connectivity of cyber-physical objects in critical environments like industrial production, public transport, or energy distribution, also increases the vulnerability of these IoT-based, cyber-physical systems. Ensuring security in such environments introduces major challenges and high effort for quality assurance. Conventional approaches for quality assurance and testing are characterized by manual tasks and script-based automation attempts, which limit the flexibility necessary for adapting to heterogeneous, dynamic and self-organizing IoT-based systems.

Adequate quality assurance and testing approaches are required to cope with a huge number and variety of interaction scenarios typical for IoT systems.

© Springer Nature Switzerland AG 2020
G. Kotsis et al. (Eds.): DEXA 2020 Workshops, CCIS 1285, pp. 48–62, 2020.
https://doi.org/10.1007/978-3-030-59028-4_5

In these scenarios, communication can run synchronously as well as asynchronously between an arbitrary number of devices acting as server and as client. Devices can be in numerous different states, triggered by the interactions with other devices or due to events in their environment. Thus, the overall configuration of the system under test is highly volatile and prone to change at run-time. In contrast, conventional automated testing aims at automating the execution of test cases that represent predefined scenarios, intended to run in a controlled and stable setting. These tests are usually created in a laborious manual step. Furthermore, whenever the setting changes, manual intervention is required, causing yet again additional effort for maintaining the existing tests.

In order to cope with the dynamic nature of IoT systems, we propose the use of testing approaches that have the ability to self-adapt to a given setting by automatically generating test cases for a running system. Several such approaches have been a subject of research in automated software testing, for example, search-based testing [2], guided random testing [16], or black-box fuzzing [15]. Hence, we developed a test architecture that allows to harnesses automated test case generation approaches for security testing of IoT systems. The architecture has been designed in alignment with key requirements from test automation [24] and provides a reusable conceptual solution for implementing different, technology-specific testing frameworks.

In this paper, we describe an implementation of the proposed architecture for automated security testing for IoT systems based on the Message Queuing Telemetry Transport (MQTT) protocol [5,6]. MQTT is among the preferred publish/subscribe protocols for IoT systems and has been implemented for a wide range of platforms and programming languages.

Despite the fact that it has been designed to be lightweight and easy to use, it still contains relatively complex protocol logic for handling connections, subscriptions, and the various quality of service levels related to message delivery. Support for security mechanisms, on the other hand, has not been a primary objective. Therefore, MQTT became a potential target for adversaries and a wide array of attack scenarios has been identified [4,9,11]. Incorporating such attack scenarios in testing is a key element of our architecture for automated security test case generation. The approach is able to generate valid interaction sequences that achieve a deep coverage of internal states of the MQTT system combined with invalid and malicious interactions that aim to expose security-relevant faults, i.e., exploitable vulnerabilities in the tested system.

The paper is organized as follows. Next section introduces the necessary background related to the MQTT protocol, (automated) test generation, and security testing. The main contributions of the paper comprise Sects. 3–6. In particular, Sect. 3 presents the proposed architecture for automated test case generation. Section 4 details the main design decisions prompted by the implementation of the test framework for security testing MQTT systems. Section 5 presents an overview of the set of attacks derived from common attack patterns and built into our test framework. Section 6 details and discusses the results of a preliminary evaluation of our architecture used for testing MQTT security. Related work is discussed in Sect. 7. Section 8 concludes our paper and discusses future work.

2   Background

2.1   Message Queuing Telemetry Transport (MQTT)

The MQTT protocol is an open, standardized [5,6], lightweight and easy to implement protocol intended to be used for communication between IoT devices. Devices in the role of clients are able to communicate with a central MQTT broker (server), exchanging messages with the broker or with each other through the broker. The typical use case for a client is to connect to a broker, taking several actions, and then to disconnect again. Actions may include publishing a message to an arbitrary topic or subscribing to one of the existing topics and, thus, receiving messages published on the topic by others. An example of such a communication sequence between a broker and its clients is depicted in Fig. 1. Overall, there exist 14 different MQTT message types including connect, disconnect, publish, subscribe, and unsubscribe. Clients are able to define the quality of service level for message delivery: *QoS 0* denotes that each packet is sent (at most) once and no retries are performed, *QoS 1* that the message arrives at the receiver at least once, and *QoS 2* that the message is delivered to the receiver exactly once.

2.2   Automated Test Case Generation

Testing is one of the most widely applied approaches for quality assurance in industry. Its practical relevance results from its ability to find faults by actually executing the system under test and triggering observable failures as evidence. However, creating test cases that have a high chance reveal faults and executing these tests on the real system is a labor-intensive and costly activity. The goal of automated test generation is to reduce the manual workload by automatically generating a set of executable test cases based on information derived e.g., from specifications or models, program structure, or the input/output observed from system executions. A variety of test generation techniques exists, including model-based testing, combinatorial testing, search-based testing, and adaptive random testing. An overview and discussion of these techniques is provided in [3].

Test cases typically consist of a sequence of interactions (like in the communication of a broker with its client), test data, and checks to assess the expected result. Predicting the expected result is one of the main challenges in automatically generating test cases. Usually it is not possible to exactly define the expected result in test generation when there is no description or model of the system available. Therefore, one has to rely on general properties such as "the system must not crash", which can be assumed to be always true and which is particularly relevant for security testing as crashes can result in exploits and denial of service attacks. Another way to determine expected results is taking the observed output recorded in a previous test run as expected output for future runs in regression testing.

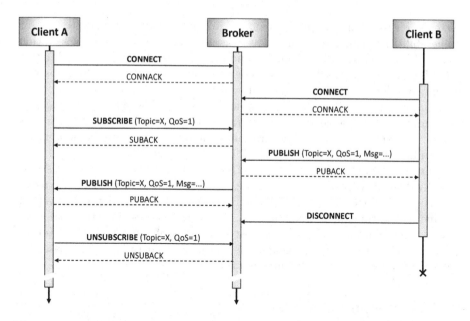

Fig. 1. Two clients connecting to a broker. Client A subscribes to topic X. When client B publishes a message (temperature = 25) to topic X, the broker distributes this value to the clients subscribed to this topic. After receiving the updated values, client A unsubscribes from topic X.

2.3   Security Testing

The aim of security testing is to reveal loopholes in software that can lead to the violation of security requirements. Typical security requirements are confidentiality, integrity and availability. Requirements can be divided into two categories [1]: positive requirements and negative requirements. The former define the expected functionality, i.e., how the system should behave. An example of a positive requirement of the MQTT protocol[1] is: "if a Client has been successfully authenticated, a server implementation should check that it is authorized before accepting its connection". The latter defines possible vulnerabilities, i.e., how the system should not behave. An example of a negative requirement with respect to the MQTT protocol would be: "a server implementation should not accept a connection of an unauthorized client". Thus, two types of security testing can be distinguished, functional security testing (positive) and security vulnerability testing (negative).

For testing positive security requirements we can use established functional testing techniques. Vulnerability testing is on the other hand more challenging, since it is usually infeasible to exhaustively check all possible cases that result in a negative outcome. This is especially true for distributed asynchronous

---

[1] See Sect. 5.4.2 at https://docs.oasis-open.org/mqtt/mqtt/v5.0/os/mqtt-v5.0-os.html#_Toc3901014.

systems consisting of multiple autonomous IoT devices. In such cases, testing for vulnerabilities either uses simulated attacks, which is also known as penetration testing [7], or fuzzing [30]. Fuzzing techniques have already been proposed to test MQTT implementations [25]. This is not surprising since the idea of fuzzing was initially developed for identifying security flaws in protocol implementations due to improper processing of malicious inputs. In our work we also target security flaws in MQTT implementations, but rather than concentrating on malicious inputs, we explore automated generation techniques to discover vulnerabilities in harmful interactions between MQTT brokers and clients.

Our approach is based on attack patterns. Attack patterns are abstractions of individual attacks into blueprints describing common methods for exploiting software vulnerabilities [13]. They are generated from in-depth analysis of specific real-world exploit examples. An attack pattern describes the goal, conditions, individual actions and the post conditions of the attack [20]. Key to our approach, attack patterns enable the automated generation of test cases for vulnerability testing (see [8, 29] among others).

3   Architecture Overview

The overall architecture consists of four core components instanced, for example, by a build and continuous integration pipeline. Figure 2 provides an overview of the architecture. The *system under test (SUT)* is the monitored IoT system which the test interacts with in order to detect security defects. An *adapter* is used to handle the communication with the SUT. A *test case generator* uses the adapter to interact with the SUT and to generate test cases in the form of possible interaction scenarios. To run the generated tests, a *test case execution framework* is applied. Overall, the architecture is designed such that it enables a transparent integration and reuse of the different components. In the following subsections the implementation of the architecture for testing MQTT-based IoT systems is described.

3.1   System Under Test (MQTT Broker)

The SUT in our case is an MQTT-based IoT system that requires at least one MQTT broker (connected to further clients) to interact with. For the purpose of demonstrating our implementation and for generating test cases we used the popular Eclipse Mosquitto[2] broker standalone. The broker simply listens to a specified port and handles incoming messages as usual. In our scenario, the SUT (MQTT broker) does not requires any specific configuration or adaptation. A monitor is used to observe and report the state of the SUT, e.g., broker status (running/idle/down), resource usage, code coverage, or crashes.

---

[2] https://mosquitto.org/.

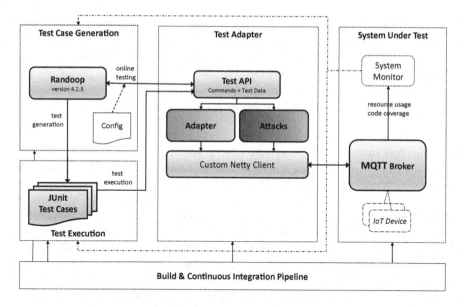

Fig. 2. Overview of the automated security test architecture.

3.2 MQTT Adapter

The adapter handles the communication between test generation/execution and the SUT. The adapter resembles an MQTT client to interact with the SUT represented by the MQTT broker (i.e., for sending MQTT messages including test data). The client exposes all possible MQTT commands such as connecting to the broker, subscribing to a channel, publishing a message, etc. in form of a Java API so they can be accessed by the test generator. When an API method from the adapter is called, it sends the corresponding message to the SUT, blocks until the SUT responds, and returns the result back to the caller. In order to also provoke negative interaction scenarios (invalid, illegal and malicious interactions) and for simulating possible attacks on the SUT, we extended the adapter API with implementations of attack patterns.

We have implemented the adapter in Java using functionality of the freely available Netty framework,[3] an asynchronous event-driven network application framework for developing server and clients using common network protocols including MQTT. It enables us to handle every aspect of the MQTT protocol, with complete control of the messages sent and received by the client (including headers, payload, low level encoding, and wire communication). This level of control is important to also construct invalid messages for implementing possible attacks. This approach also enables us to handle multiple concurrent client connections. We structured the adapter in three layers. The bottom layer is a customized Netty client, the middle layer provides a set of methods to

---

[3] https://netty.io.

conveniently work with the low-level Netty client, and the top layer consists of a set of methods for each MQTT message type with parameters set to predefined test data in various combinations, so they can be called by the test generator without the need to generate adequate test data.

3.3   Test Case Generation (Randoop)

For generating test cases we applied the open source tool Randoop [23], a feedback directed random test generator that generates random but viable interaction sequences by exploring the SUT trough random calls to its public API. Randoop supports testing Java APIs without the need of additional input, so it can be run directly on the Java test adapter we developed. Furthermore, the generated test case will also include invalid and illegal command sequences, since the implemented attack pattern are also available in form of API method of the test adapter. Hence, the test case generator is able to create interaction sequences that consist of valid sequences interleaved by malicious interactions. This interleaving allows to bring the SUT into a specific state before performing an attack. Many attacks have preconditions, for example, they require a successfully initiated connection and the subscription to an existing topic.

```
@Test
public void test01 () throws Throwable {

    /* Creating Client A and connecting to broker */
    MqttClientAdapter clientA = new MqttClientAdapter ();
    String str1 = clientA.connectQoS0 ();
    assertTrue ( str1 . equals ("MqttConnAck [/* ... */]"));

    /* Creating Client B and connecting to broker */
    MqttClientAdapter clientB = new MqttClientAdapter ();
    String str2 = clientB.connectQoS0 ();
    assertTrue ( str2 . equals ("MqttConnAck [/* ... */]"));

    /* Client A subscribing to topic X */
    String str3 = clientA.subscribeIntervallTopicXQoS1 ();
    assertTrue ( str3 . equals ("MqttSubAck [/* ... */]"));

    /* Client B publishing to topic X and disconnecting */
    MqttMsgId mqttMsgId1 = clientB.publishIntervallTopicXQoS1 ()
    assertNotNull ( mqttMsgId1 );
    clientB.disconnectQoS1 ();

    /* Client A receiving message and unsubscribing */
    String str4 = clientA.unsubscribeIntervallTopicQoS1 ();
    assertTrue ( str4 . equals ("MqttUnsubAck [/* ... */]"));
}
```

Listing 1.1. Example JUnit test case generated with Randoop

3.4 Test Case Execution (JUnit)

Randoop outputs the generated test sequences as JUnit test cases, so that they can be executed in regression testing without the need to re-run the test generator. JUnit is the most popular and widely used framework for unit testing and it is supported by all major Java development tools, which comes with the benefit that the generated tests can be directly integrated in an existing build and continuous integration pipeline.

An example for a JUnit test case generated by Randoop is presented in Listing 1.1. The example shows the interaction of two clients A and B with an MQTT broker following the scenario illustrated in Fig. 1 above. The clients use our MQTT test adapter to send messages to the broker and the corresponding response from the broker (e.g., acknowledgement messages) are checked using JUnit's assert methods.

4 Design Decisions

During the implementation phase of our prototype we had to address several technical challenges. In this section we briefly describe the main design decisions prompted by these challenges.

1. **Risk-based attack patterns integration.** In [18] we describe a general approach to automatically test security of industrial IoT systems. In that approach we identify possible attack patterns through means of threat modelling and risk assessment. Based on those patterns we generate actual attacks adapted to the particularities of the SUT, which are then run periodically. We integrate this idea into our architecture by adding the possibility of defining attack methods within the adapter component. These attacks are then used during the automated test generation phase.
2. **Customized "raw" MQTT client.** Standard MQTT clients perform a series of checks to make sure that the messages they handle are in line with the MQTT standard. For example, the payload size parameter is automatically corrected to fit the message sent to the broker. This means that standard MQTT clients are not suitable to test invalid scenarios such as corrupted payload values or invalid message encoding. To deal with this issue, we developed our own customized MQTT client and encoder/decoder components based on the Netty framework. They provide a fluent "raw" interface to interact with MQTT brokers that allows us to test arbitrary malicious attacks.
3. **Regression tests.** One of our aims is to detect discrepancies among different implementations of MQTT brokers, which can indicate bugs and potential security threats. Thus, we designed our architecture so that it can be used for regression testing. In particular, the test generation component can store the generated test sequences. These test sequences can then be used by the test execution component to re-execute the tests against diverse MQTT broker implementations, storing the results of each executed test replication. Finally, corresponding results are pairwise compared against each other in order to detect discrepancies in the behaviour of different MQTT brokers.

4. **Clean broker state.** We decided to restart the SUT before each test generation as well as before each test sequence execution, so that each time we start with a clean system state. This is necessary to ensure consistent deterministic behaviour of the test executions in situations in which clients created by different independent test sequences share the same identifier. Note that an MQTT client can connect with the clean-session flag turned off (see MQTT v3.1.1 standard [6]) or with a long session expiry interval (see MQTT v5.0 standard [5]), and thus receive for instance messages from a different client from a previous test run with a same client identifier - which is of course an undesirable side effect.

5. **Monitoring broker state.** During test generation and execution we need to constantly monitor the state of the MQTT broker. In particular, we need to know whether the broker operates normally. Abnormal broker behaviour such as unresponsiveness due to extremely high resource usage indicates that our test sequence might have found some vulnerability in the broker. If the broker is not operational, we need to know as testing should then be aborted. In some cases it is also necessary to gather additional information (e.g. code coverage [21]) in order to apply advanced test generation and execution techniques.

6. **Support for multiple clients.** In order to fully test MQTT systems, it is of course necessary to handle multiple, concurrent client connections to MQTT brokers. In our architecture this is supported by the adapter component. We have to be careful however with certain details, in particular when long test sequences are generated. If left unchecked such long test sequences can lead to performance problems during test generation and execution. For instance, they might include way too many client subscriptions to a given topic, which in turn can lead to performance issues due to the overheads resulting of distributing MQTT messages submitted to this topic to such a big number of subscribers. Our test generation module takes care of these problematic cases by flagging such anomalous test sequences.

5   Attack Patterns

In our prototype we implemented a set of predefined attacks. These attacks are derived from attack patterns described in collections such as Common Attack Pattern Enumeration and Classification Database (CAPEC). We derived the actual attacks manually, analysing common attach patterns and adapting them to the MQTT case. In addition to documented attack patterns for vulnerability testing, we have devised attacks also for functional security. As expected, the latter were defined based in the specifications of the MQTT standards [5,6]. Table 1 presents a short description of the attacks that we have been able to define using this approach. Most of these attacks have already been implemented in our framework, except for those marked with an asterisk.

In order to give the reader a better insight on how these attacks actually work, we briefly discuss selected attacks.

Table 1. Overview of possible attacks.

Id	Short description	CAPEC	MQTT
1	Sends two identical CONNECT packets		x
2	Connects with empty client identifier	x	
3	Connects with a non-safe form of UTF8 encoding		x
4	Connects with invalid protocol id and/or version[a]		x
5	Connects with invalid QoS		x
6	Connects with invalid protocol level[a]		x
7	Connects with invalid reserved flag		x
8	Connects with invalid will flags		x
9	Connects with invalid user password flags	x	
10	Connects with invalid user name	x	
11	Connects with payload values in wrong order		x
12	Connects with high keep alive values	x	
13	Connects with invalid client id		x
14	Connects with long client id	x	
15	Subscribes with missing payload	x	
16	Subscribes with invalid reserved flag[a]		x
17	Subscribes with invalid topic filter	x	
18	Subscribes with wildcards in topic name		x
19	Subscribes with escape sequences in topic	x	
20	Subscribes with deep topic hierarchy depth	x	
21	Subscribes without giving a topic name	x	
22	Unsubscribe without giving a topic name	x	
23	Publishes with escape sequences in payload	x	
24	Publishes a very big payload (128 MB)	x	
25	Publishes with wildcards in payload		x
26	Publishes with both QoS-Bits set (QoS value is 3)[a]		x
27	Publishes without giving a message	x	
28	Publishes with empty topic	x	
29	Publishes with invalid variable header length	x	

[a] Implementation in progress.

Attack 1: This attack tests how the broker behaves if a same client sends two CONNECT packets to the broker. A Client can only send the CONNECT packet once over a network connection. The broker must treat a second CONNECT packet sent from a client as a protocol violation and disconnect the client (see [MQTT-3.1.0-2] in [6]). While the Mosquitto broker that we used for evaluation in this paper proceeds correctly under this attack, it is known that other brokers such as HBMQTT (version 0.7.1) does not, since it simply ignores a second connection attempt [31].

Attack 20: This attack tries to provoke an stack overflow by sending subscription messages to levels too deep in the topics' hierarchy. It is done by sending subscriptions to topics with 65400 or more topic hierarchy separators '/'. As mentioned in the introduction, it is known that this attack will provoke an stack overflow in the mosquitto MQTT brokers version 1.5.0 to 1.6.5 inclusive.

Attack 29: The variable part of a publish message can contain several message fields. Their lengths depend on the type of the field and the actual payload. For example, we can publish an arbitrary message to a given topic. The size of the variable header part has to be given accordingly. In this attack we deliberately change the specified length of the variable header to an incorrect value in order to trigger faults in the handling of the packet by the MQTT-Broker which can lead to a buffer overflow.

6 Evaluation

In this section we present and discuss the results of a preliminary evaluation of our implementation for security testing of MQTT systems. More specifically, we evaluate the overall performance by measuring the coverage achieved by the generated tests. Besides running the tests on the widely used MQTT broker Mosquitto we used the Java-based open source MQTT broker Moquette[4] for this preliminary evaluation.

Table 2 shows the results obtained from this initial experiment. In the first column the package structure of the broker implementation is depicted. For reference, the second column in the table shows the code coverage attained by simply starting and stopping the broker again. The third column shows the code coverage attained by executing the generated tests via our adapter. We performed the experiments for test case generation based on Randoop for 10 cycles with each running for 90 s. The values in the corresponding column show the average code coverage results. Finally, the fourth column in the table shows the average code coverage attained by 10 runs with the same preconditions as before, but including the implemented attacks from Table 1.

The increase in code coverage obtained when attacks derived from attack patterns are included in the test generation process appears not to be significant. However, the increase indicates successful execution of the attacks on the broker and one must note that for many attacks the broker will not return an answer. This leads to an increase in the number of timeouts, lowering the overall number of messages sent to the broker in the time span of 90 s of each experiment. This means that by including attacks in the test generation process, we already were able to maintain (even with a small increase) the code coverage while reducing the overall number of MQTT packets interchanged with broker.

---

[4] https://github.com/moquette-io/moquette.

Table 2. Code coverage measurements.

Package	Start-stop	Generated	Attacks
io.moquette	0.0	0.0	0.0
io.moquette.broker	0.13	0.516	0.526
io.moquette.broker.config	0.49	0.49	0.49
io.moquette.broker.metrics	0.04	0.762	0.762
io.moquette.broker.security	0.10	0.11	0.118
io.moquette.broker.subscriptions	0.06	0.529	0.548
io.moquette.interception	0.15	0.316	0.316
io.moquette.interception.messages	0.0	0.0	0.0
io.moquette.logging	0.62	0.62	0.62
io.moquette.persistence	0.02	0.079	0.074
Total	0.12	0.456	0.463

7   Related Work

Research on secure MQTT systems is still scattered. The protocol itself is not concerned with security, though it provides a basic mean to control who can send and receive messages through user name and password, which are included in the MQTT packets.

Some research such as [12] has concentrated on adding security layers around the TCP/IP protocol, identifying security architectures and models that best fit IoT networks. Others have designed and propose frameworks to secure the transport layer by using SSL/TLS [17,26], trying to minimize the significant network overheads. Works such as [27,28] have focused in lightweight encryption approaches to secure MQTT communication channels. This is an acute problem when dealing with smart devises that do not have enough processing power to use asymmetric encryption algorithms for authentication tasks. Thus, alternatives for lightweight and secure authentication have been proposed [10].

In this paper we are instead concerned with behavioural faults triggered by the iteration between brokers and clients working concurrently in a production environment, which can lead to unforeseen security breaches. The most trivial of such faults are due to the state in which the protocol works by default. Since MQTT is a simple protocol designed for devices with low processing power, the default configuration aims to minimize the processing needed to exchange messages rather than at security which means that serious security problems arise. Most of these shortcomings can be solved with an adequate protocol configuration.[5] A stronger approach in this sense is to enforce compliance with security features that go beyond the default MQTT implementation, which can be done through a model-based security toolkit [22].

---

[5] See for instance https://www.hivemq.com/downloads/hivemq-data-sheet-4.2.pdf.

In any case, and independently of the diverse approaches follow to secure MQTT implementations, we need tools to test how well they work in practice, detecting security holes before they become a problem. With this aim, a framework to test security of applications that implement MQTT has been proposed in [25]. This framework enables the application of template-based fuzzing techniques to the MQTT protocol. Our approach here is complementary with that, since we propose a precise architecture for applying random testing, rather than a framework for applying fuzzing testing techniques. As future work, it would indeed be interesting to explore possible combinations of both approaches.

Model based formal testing has also been used for MQTT, but with the aim of checking the conformance to the standard of different MQTT brokers [14,19]. In [31] a learning-based approach to detecting failures in reactive systems was applied to obtain an automata modeling the communication with MQTT brokers, which was then used to detect discrepancies between different implementations. Several bugs were discovered in popular implementations of MQTT browsers. Although interesting, this method has also its limitations due to the difficulty of automata learning under non-deterministic concurrent executions.

8 Conclusions and Future Work

We have proposed an architecture for automated generation of security tests for IoT systems and we described an implementation of the proposed architecture for test case generation specifically for systems based on the MQTT protocol. The implementation shows how to realize the different components of our architecture, for example, we incorporated guided random testing using the open source tool Randoop for test case generation. Moreover, our implementation also employs attacks derived from common attack patterns resembling possible malicious interactions to perform vulnerability testing in addition to functional security testing. Based on our implementation we discussed the key technical challenges we had to solve and we performed an initial evaluation using an open source MQTT broker.

Although the results obtained so far are very encouraging, this is still a work in progress. In order to confirm our initial results and to exploit the full potential of the proposed architecture for security testing, we plan following future work:

- Increase substantially the number of attacks by means of an exhaustive analysis of attack patterns obtained from publicly available catalogs.
- Produce a formal threat model as sketched in [18] and use it to automate the task of selecting meaningful security tests based in sound risk assessments.
- Perform regression tests considering different implementations of MQTT brokers and different test environments, analyse the discrepancies found via this method and classify them according to a risk based assessment.
- Perform a deep analysis of threat coverage of the proposed architecture. The preliminary code coverage we have performed so far, is just part of the equation, since there is no clear correlation between code coverage and effectiveness of the security test iteration sequences.

- Integrate the template-based fuzzing techniques used in [25] to test security of the MQTT protocol into the proposed architecture. Perform experiments to determine if this leads to an increase in number of reported security threads.

Acknowledgement. The research reported in this paper has been supported by the *ICT of the Future* programme (grant #863129, *IoT4CPS*) and the *COMET Competence Centers Programme* (grant #865891, *SCCH*) managed by FFG and funded by the Austrian federal ministries BMK and BMDW, and the Province of Upper Austria.

References

1. Alexander, I.F.: Misuse cases: use cases with hostile intent. IEEE Softw. **20**(1), 58–66 (2003)
2. Ali, S., Briand, L.C., Hemmati, H., Panesar-Walawege, R.K.: A systematic review of the application and empirical investigation of search-based test case generation. IEEE Trans. Softw. Eng. **36**(6), 742–762 (2009)
3. Anand, S., et al.: An orchestrated survey of methodologies for automated software test case generation. J. Syst. Softw. **86**(8), 1978–2001 (2013)
4. Andy, S., Rahardjo, B., Hanindhito, B.: Attack scenarios and security analysis of MQTT communication protocol in IoT system. In: 4th International Conference on Electrical Engineering, Computer Science and Informatics (EECSI), pp. 1–6. IEEE (2017)
5. Banks, A., Briggs, E., Borgendale, K., Gupta, R.: MQTT Version 5.0. OASIS Standard. https://docs.oasis-open.org/mqtt/mqtt/v5.0/mqtt-v5.0.html
6. Banks, A., Gupta, R.: MQTT Version 3.1.1. OASIS Standard. http://docs.oasis-open.org/mqtt/mqtt/v3.1.1/mqtt-v3.1.1.html
7. Bishop, M.: About penetration testing. IEEE Secur. Priv. **5**(6), 84–87 (2007)
8. Bozic, J., Wotawa, F.: Security testing based on attack patterns. In: Seventh IEEE International Conference on Software Testing, Verification and Validation, ICST 2014 Workshops Proceedings, 31 March–4 April 2014, Cleveland, Ohio, USA, pp. 4–11. IEEE Computer Society (2014)
9. Dinculeană, D., Cheng, X.: Vulnerabilities and limitations of MQTT protocol used between IoT devices. Appl. Sci. **9**(5), 848 (2019)
10. Esfahani, A., et al.: A lightweight authentication mechanism for M2M communications in industrial IoT environment. IEEE Internet Things J. **6**(1), 288–296 (2019)
11. Firdous, S.N., Baig, Z., Valli, C., Ibrahim, A.: Modelling and evaluation of malicious attacks against the IoT MQTT protocol. In: IEEE International Conference on Internet of Things (iThings) and Green Computing and Communications (GreenCom) and Cyber, Physical and Social Computing (CPSCom) and IEEE Smart Data (SmartData), pp. 748–755. IEEE (2017)
12. Heer, T., Morchon, O.G., Hummen, R., Keoh, S.L., Kumar, S.S., Wehrle, K.: Security challenges in the IP-based internet of things. Wireless Pers. Commun. **61**(3), 527–542 (2011). https://doi.org/10.1007/s11277-011-0385-5
13. Hoglund, G., McGraw, G.: Exploiting Software: How to Break Code. Addison Wesley, Boston (2004)
14. Houimli, M., Kahloul, L., Benaoun, S.: Formal specification, verification and evaluation of the MQTT protocol in the internet of things. In: 2017 International Conference on Mathematics and Information Technology (ICMIT), pp. 214–221. IEEE Computer Society (2017)

15. Liang, H., Pei, X., Jia, X., Shen, W., Zhang, J.: Fuzzing: state of the art. IEEE Trans. Reliab. **67**(3), 1199–1218 (2018)
16. Ma, L., Artho, C., Zhang, C., Sato, H., Gmeiner, J., Ramler, R.: GRT: program-analysis-guided random testing (T). In: 2015 30th IEEE/ACM International Conference on Automated Software Engineering (ASE), pp. 212–223. IEEE (2015)
17. Manzoor, A.: Securing device connectivity in the industrial Internet of Things (IoT). In: Mahmood, Z. (ed.) Connectivity Frameworks for Smart Devices. CCN, pp. 3–22. Springer, Cham (2016). https://doi.org/10.1007/978-3-319-33124-9_1
18. Marksteiner, S., Ramler, R., Sochor, H.: Integrating threat modeling and automated test case generation into industrialized software security testing. In: Proceedings of the Third Central European Cybersecurity Conference, CECC 2019, Munich, Germany, 14–15 November 2019, pp. 25:1–25:3. ACM (2019)
19. Mladenov, K.: Formal verification of the implementation of the MQTT protocol in IoT devices. Technical report, University of Amsterdam, Faculty of Physics, Mathematics and Informatics (2017)
20. Moore, A., Ellison, R., Linger, R.: Attack modeling for information security and survivability. Technical report, Technical Note CMU/SEI-2001-TN-001, Carnegie Mellon University (2001)
21. Nagy, S., Hicks, M.: Full-speed fuzzing: reducing fuzzing overhead through coverage-guided tracing. In: 2019 IEEE Symposium on Security and Privacy (SP), pp. 787–802 (2019)
22. Neisse, R., Steri, G., Baldini, G.: Enforcement of security policy rules for the Internet of Things. In: IEEE 10th International Conference on Wireless and Mobile Computing, Networking and Communications, WiMob 2014, pp. 165–172. IEEE Computer Society (2014)
23. Pacheco, C., Ernst, M.D.: Randoop: feedback-directed random testing for java. In: Companion to the 22nd ACM SIGPLAN Conference on Object-Oriented Programming Systems and Applications Companion, pp. 815–816 (2007)
24. Ramler, R., Buchgeher, G., Klammer, C.: Adapting automated test generation to GUI testing of industry applications. Inf. Softw. Technol. **93**, 248–263 (2018)
25. Ramos, S.H., Villalba, M.T., Lacuesta, R.: MQTT security: a novel fuzzing approach. Wireless Communications and Mobile Computing **2018** (2018)
26. Shin, S., Kobara, K., Chuang, C., Huang, W.: A security framework for MQTT. In: 2016 IEEE Conference on Communications and Network Security, CNS 2016, Philadelphia, PA, USA, 17–19 October 2016, pp. 432–436. IEEE (2016)
27. Singh, M., Rajan, M.A., Shivraj, V.L., Balamuralidhar, P.: Secure MQTT for Internet of Things (IoT). In: Fifth International Conference on Communication Systems and Network Technologies, pp. 746–751. IEEE (2015)
28. Su, W., Chen, W., Chen, C.: An extensible and transparent thing-to-thing security enhancement for MQTT protocol in IotTenvironment. In: 2019 Global IoT Summit, GIoTS 2019, Aarhus, Denmark, 17–21 June 2019, pp. 1–4. IEEE (2019)
29. Sudhodanan, A., Armando, A., Carbone, R., Compagna, L.: Attack patterns for black-box security testing of multi-party web applications. In: 23rd Network and Distributed System Security Symposium, NDSS 2016, San Diego, CA, 21–24 February 2016. The Internet Society (2016)
30. Takanen, A., DeMott, J., Miller, C.: Fuzzing for Software Security Testing and Quality Assurance, 1st edn. Artech House, Inc., Norwood (2008)
31. Tappler, M., Aichernig, B.K., Bloem, R.: Model-based testing IoT communication via active automata learning. In: IEEE International Conference on Software Testing, Verification and Validation, ICST 2017, Tokyo, Japan, March 2017, pp. 276–287. IEEE (2017)

Mode Switching from a Security Perspective: First Findings of a Systematic Literature Review

Michael Riegler$^{(\boxtimes)}$ and Johannes Sametinger

Department of Business Informatics, LIT Secure and Correct Systems Lab,
Johannes Kepler University, Linz, Austria
{michael.riegler,johannes.sametinger}@jku.at
https://www.jku.at/en/lit-secure-and-correct-systems-lab

Abstract. With increased interoperability of cyber-physical systems (CPSs), security becomes increasingly critical for many of these systems. We know mode switching from domains like aviation and automotive, and we imagine to use this mechanism for the development of resilient systems that continue to function correctly even if under malicious attack. If vulnerabilities are detected or even known, modes can be switched to reduce the attack surface and to minimize attackers' range of activity. We propose to engineer CPSs with multi-modal software architectures to overcome the interval between the time when zero-day vulnerabilities become known and the time when corresponding updates become available. Thus, affected companies, operators and people will be able to protect themselves and their customers without having to wait for security updates. This paper presents first findings of a systematic literature review (SLR) on mode switching from a security perspective.

Keywords: Mode switching · Protocols · Security · Resilience · Cyber-physical system

1 Introduction

Cyber-physical systems (CPSs) are based on real-time embedded devices with wired or wireless connections to monitor and control physical systems and are used in mission- or safety critical applications like aviation, self-driving cars, unmanned aerial vehicle, medical monitoring, traffic control, power plants and industrial machines. Such systems require functional correct execution on time.

A CPS consists of hardware and software that interacts with physical components like sensors and actuators and with other local and/or remote systems. To master complex systems they are divided into logical, controllable and tangible modes of operation. Each *mode* has different goals as well as its unique behavior, and it is characterized by a set of functionalities. Such a multi-modal architecture makes a system more flexible to respond to external events. For example

© Springer Nature Switzerland AG 2020
G. Kotsis et al. (Eds.): DEXA 2020 Workshops, CCIS 1285, pp. 63–73, 2020.
https://doi.org/10.1007/978-3-030-59028-4_6

airplanes have modes for parking, taxiing, take-off, manual and automatic cruising flight, landing and possibly emergency. A *mode switch* is used to change the behavior, e.g., from take-off to cruising flight or from cruising flight to landing.

An increasing number of CPSs goes hand in hand with security vulnerabilities and potential threats to these devices. As CPSs consists of many hardware and software components each of these building blocks can be an attack gateway. To resolve such vulnerabilities, manufacturers provide updates and patches. But developing and distributing them may take a long time particularly with regard to safety requirements. Meanwhile, attackers can write exploits to take advantage of the vulnerabilities. We are targeting security at the application layer and propose a mode switching mechanism as an effective countermeasure to cyber-security threats that is based on [29]. We can switch modes automatically based on a kind of intrusion detection system or manually by an operator. Thereby, involved and affected parties get a tool to protect themselves to reduce attack surface and limit attackers' range of activity.

In this paper, we present first findings of a systematic literature review (SLR) on mode switching from a security perspective. We identify and analyze the current state of mode switching and how it can be used for security. Our focus is on proposed protocols and techniques. We intend to adapt and combine existing approaches from other domains to increase security. In Sect. 2, we describe mode switching in general and from a security perspective in Sect. 3. In Sect. 4, we introduce our SLR and present the first results of our review. In Sect. 5, we forge a bridge back to security and safety, and draw our conclusions in Sect. 6.

2   Mode Switching

The idea of *multiple modes* and *mode switching* is well known from domains like aviation (parking mode, taxiing mode, take-off mode, manual and automatic cruising flight mode, landing mode) and automotive (manual mode, adaptive cruise control mode, lane keeping assistant mode, emergency braking mode, parking mode) [8,28]. Such multi-mode systems are also used to manage complexity and to divide systems into modes of operation in the automotive domain.

Real and Crespo [32] define a mode switch as a transient stage between the current operating mode, the old mode, to another mode, the new mode. It is used to change the behavior and performance [19]. Modes can be organized hierarchically and by having several sub-modes. The functionality of each mode consists of a series of tasks. *Tasks* can run periodically or sporadically, some will run independently in the background, and some of them may only be necessary within specific modes.

A mode switch is initiated by a mode switch request (MSR) or mode change request (MCR) and is triggered by a *timer* or a specific *event*. For example, a smart phone poll mail servers every 15 min (timer) and go into an idle or sleeping mode to save battery thereafter. When a phone call comes in (event), an active mode takes over from the idle mode. During a mode change, continuous and mode-independent tasks should not be interrupted. New tasks are released and no longer needed tasks are stopped. Modes may execute both old-mode and new-mode tasks. The scheduler has to guarantee deadlines and prevent an overload of the system.

3   Security Modes

Security is about protecting information and information systems from unauthorized access and use. Confidentiality, integrity and availability of information are the core goals. Software security is "the idea of engineering software so that it continues to function correctly under malicious attack" [20].

Systems are considered to be resilient if they do their work despite adverse circumstances and if they recover rapidly from any harm [9]. No matter how well systems have been designed, unknown or unresolved vulnerabilities can become a risk in the future. Resilient systems protect their important assets by using methods to detect adversities, to mitigate them and to recover. For resilient security, exposure and authentication measures may be adapted depending on the security context by switching modes [34].

But while mode switching has been known from various domains, as to our knowledge, it has never been used in the domain of security. This fact has motivated us to do a systematic literature review as a first step in doing just that, i.e., making systems more resilient, engineering systems so that they continue to function correctly under malicious attack. We imagine the use of multiple modes for the purpose of increasing security and making systems more resilient.

4   Systematic Literature Review

Guidelines for conducting a systematic literature review (SLR) have been presented in [16]. We have made a preliminary search to verify that no recent SLRs about mode switching have been done. This verification was performed by searching through different databases (ACM Digital Library, IEEE Explore and Science Direct) with terms related to the methodology ("SLR", "systematic literature review", etc.) and the target of the review ("mode switching, "mode change", etc.). A survey on mode change protocols for single-processor, fixed-priority, preemptively scheduled real-time systems has been presented in [32], but the results of these queries confirmed that there are no previous SLRs about mode switching.

Based on the PICOC method proposed by [27], we have defined the review scope as follows:

- **Population (P):** Software or System
- **Intervention(I):** Approaches, Protocols, Frameworks, Methods and Mechanisms
- **Comparison (C):** The primary goal of the SLR is to analyze and compare existing approaches and gain knowledge about them
- **Outcomes (O):** Mode switching
- **Context(C):** Security perspective

We want to address the following research questions (RQ):

- **RQ1.** Which protocols for mode-switching are currently available?
- **RQ2.** For what purposes has mode-switching been used?
- **RQ3.** How do switches from one mode to another really work? What are the pitfalls?
- **RQ4.** Are there approaches which use mode-switching from a security perspective?

Based on the review scope we defined the search term ("mode switching", "mode change", and its synonyms) AND ("framework" OR "mechanism" OR "method" OR "protocol", etc.). ACM, IEEE and ScienceDirect have been chosen for our review, because they publish a substantial amount of peer-reviewed papers on the subject of computer science. Conducting an SLR is a tedious and time-consuming task. We have not yet finished, but will present first important findings in this paper. More insights will be presented when the SLR is finished. In our pilot study a total of 1,804 raw papers were retrieved: 42 from ACM, 495 from IEEE and 1,267 from ScienceDirect. After organizing the papers in a spreadsheet 956 duplicate works were removed and 848 papers remained.

First results of our literature review have shown various aspects of mode switching. We will briefly introduce them in the following subsections: conditions and requirements in Sect. 4.1, protocols for single and multi-core platforms in Sect. 4.2, component-base multi-mode systems in Sect. 4.3, and programming and specification languages in Sect. 4.4.

4.1   Conditions and Requirements

A mode switching protocol has to define how a mode switch can be triggered and how this has to be requested. This can be done manually, e.g., a driver's request to take back control of the vehicle from self-driving mode, or automatically by the system itself. Several conditions can trigger mode switches.

Many CPSs and real-time systems have limited resources, and therefore it is necessary to prevent resource bottlenecks. Processors are getting faster every year, but the tasks are also growing. Dynamic resource allocation, predictability analysis and other techniques make sure to avoid bottlenecks [1,28]. Such mode switching protocols have to provide quality of service and efficient resource management [1,28,38]. In this context, a mode change can be used for energy efficiency. For example, if batteries of self-driving cars come close to empty, operation will be limited to the core system only.

Safety issues may also lead to mode switches. In mixed-critical systems (MCS), modes and tasks can have a different level of criticality or Safety Integrity Level (SIL). For example the engine electronics and the emergency braking mode of a car are more critical than the lane keeping mode and other driver assistance systems. To avoid accidents, vehicles have to promptly adapt their behavior. Nevertheless, in case of a crash, the airbag should deploy quickly to prevent the driver from hitting the dashboard. The entertainment system has a lower criticality than airbag deployment, yet all modes coexist in one "mixed critical" system. Mission critical tasks have a higher priority and must be continued even if errors or system failures occurs. Examples of such fault-tolerant designs are proposed by [6–8,39].

Considering *security* risks for a mode switch is quite new [34]. If a threat is detected, e.g., by kind of intrusion detection system, a mode switch is necessary. If a vulnerability becomes known, an operator can manually switch modes to reduce the attack surface. We imagine a middleware as software layer on top of the operating system that provides interoperability, functionality for different software components and applications and last but not least security. Resilience and fault tolerance play a major role here. Especially life-critical systems should not break down by a single error. Possibly there is a mode switch to a lower mode, but the system should continue its mission-critical operation.

4.2 Protocols for Single- and Multi-core Platforms

Real and Crespo [32] have examined mode change protocols (MCPs) for single-processor, fixed-priority, preemptively scheduled real-time systems. They structured MCPs in synchronous and asynchronous protocols and classified them based on the requirements for a mode change: schedulability, periodicity, promptness and consistency.

Real-time systems have to complete tasks within given deadlines. Schedulers have to guarantee these deadlines and, at the same time, have to prevent system overloads. A protocol specifies when new tasks are released and what happens with the unfinished tasks of the old mode. It defines whether new tasks have to wait until the old ones are finished or will run in parallel. It also defines whether the old tasks are completed or stopped immediately. Another requested mode switch during an active mode switch is another challenge. Mechanisms for conflict handling are needed.

With *synchronous* protocols, new-mode tasks are only started when the old-mode tasks have completed, and they are suitable if promptness and periodicity are not critical. Examples are the *Idle Time Protocol* [40,42], where old-mode tasks are suspended at the first idle time-instant, and single offset protocols which delay the start of all the new-mode tasks with an offset. The *Maximum-Period Offset Protocol* [3] supports periodicity and the delay of the first activation is equal with the period of the less frequent task in both modes. The *Minimum Single Offset Protocol* [31] finishes the last activation of all old-mode tasks and thereafter releases the new-mode tasks. This protocol is proposed with and without periodicity. In this context, periodicity means that the protocol is capable of dealing with mode-independent tasks.

Asynchronous protocols provide lower mode change latency but require a special schedulability analysis. These protocols allow to the activation of new-mode tasks while old-mode tasks are still running and are better suited for promptness. One example is the *Utilization Based Protocol* [37], which considers periodic tasks and shares processor capacity with a priority ceiling protocol [36]. This protocol for Fixed-Task-Priority (FTP) schedulers was extended to

Earliest-Deadline-First (EDF) by [2]. But the protocol of [37] was considered as insufficient by [41], and they proposed a protocol with priority that allows tasks to modify their parameters during mode changes. Further protocols without periodicity were presented in [25,26].

Scheduling mutli-mode applications in heterogeneous multi-processor platforms is more complex than on single-processor platforms and cannot be solved by similar techniques. There are scheduling anomalies upon multiprocessors. Several protocols have been extended to multi-processor platforms. Based on [32], Nelis et al. propose two protocols without periodicity for identical multiprocessor platforms and extended it for uniform platforms in [21,23]: The *Synchronous Multiprocessor Minimal Single Offset (SM-MSO)* protocol and the *Asynchronous Multiprocessor Minimal Single Offset (AM-MSO)* protocol. To provide periodicity, the *Synchronous Multiprocessor Maximum Deadline Offset (SM-MDO)* protocol was developed [22]. It is similar to [4], where tasks are reweighted at runtime, and uses the EDF algorithm for scheduling the modes on identical multiprocessors. For supporting mode-independent tasks, it is necessary to perform a schedulability test on the whole system.

4.3   Component-Based Multi-Mode Systems

Hang and Hansson combine Component-Based Software Engineering (CBSE) and multi-mode systems and propose a *Mode Switch Logic (MSL)* [10]. Usually, this is contradictory, because the splitting of a system into modes is a top-down approach and CBSE is a bottom-up approach. Nevertheless, they show how it is feasible to combine and integrate them. In the development of Component-Based Multi-Mode Systems, the behavior of the system is divided into several modes with reusable software components to manage complexity. To switch modes in case of an emergency, they propose an *Immediate Handling with Buffering (IHB)* to switch modes quickly.

Mode switching is used in a range of component models: Koala [24], SaveCCM [12], COMDES-II [15], RubusCMv3 [11], MyCCM-HI [5].

4.4   Programming and Specification Languages

The standardized *Architecture Analysis and Design Language (AADL)* is used in the field of automotive and avionics. It presents modes as states, which are integrated into a state machine.

Giotto is a programming language with time-triggered semantics. It is used for safety-critical applications with hard real-time constraints, because the timing behavior is highly predictable [13].

The *Timing Definition Language (TDL)* [33] builds on the concept of Giotto and is used for describing time behavior of component-based real-time applications. It provides an abstraction for independent execution times of periodic time-triggered computational task. The language allows simulations without having to know all physical platform details (CPU, memory, ...).

The programming language *Lustre* has been extended with *Mode-Automata* [18]. Maraninchi and Rémond provide a new construct which supports direct expression of mode-structure of complex systems. They define a mode as collection of execution states [17]. These concrete states are related to the program memory during execution. A sequence of modes is the normal system behavior.

In the area of *Service Oriented Computing (SOC)*, Hirsch et al. propose modes as new element of architectural descriptions, especially for complex service oriented systems [14]. They extend the *Darwin* architectural language for the description and verification of complex adaptable systems in the automotive domain. Thereby, modes help as scenario-based abstraction with service configuration to close the gap between requirements and software architectures.

5 Security and Safety

As mentioned in Sect. 4.1, using security risks as triggers for mode switches is a new approach [34]. In many CPSs, safety is more important than security. But security vulnerabilities can undermine safety. From a security perspective, patches and updates should be provided and installed as quickly as possible. Moreover, systems should react immediately (fail-secure) if antivirus software detects malicious files, or if an intrusion detection system identifies suspicious processes. From a safety perspective, any change needs an impact assessment, complete test evidence and a formal approval. Therefore, updates are rare and need long planning. Safety-relevant processes must never be interrupted or stopped without human interaction (fail-safe). Multi-modal software architectures can overcome this contradiction and lead to resilient systems. If security is considered from the very beginning of system design, it easier to engineer secure and resilient systems that works correctly as expected even during attacks.

In the development of safety critical Java applications mission modes provide different functionality without a restart [35]. Imagine a pacemaker with three modes, a normal, a surveillance, and a configuration mode. The mission goal is that several tasks (like monitoring the heart rhythm and pacing on demand) will run during all modes and should not be disturbed by mode switching. Other tasks are only needed in specific modes, e.g., sending patient and device status to a remote monitor and receiving configuration changes. These tasks have a lower priority than vital critical tasks.

Another approach from [43] is to learn modes with machine learning. Systems have a specific behavior under normal circumstances. In case of attacks like sensor spoofing or Denial of Service (DoS), system behavior changes and thereby abnormality can be detected. This approach is similar to the run-time security detector by [30] which considers run-time changes as security incident. Based on expert defined risk level the current mode is switched when a certain threshold is reached.

6   Conclusion

The idea of multi-modal systems and mode switching are used to provide schedulability, periodicity, promptness and consistency for CPSs. But we can use this approach not only to prevent resource bottlenecks, but also to engineer resilient and secure systems with a high demand on safety. Modes are part of programming and specification languages and are applied in single- and multi-core platforms to provide real-time adaptivity. In Component-Based Multi-Mode Systems reusable software components and modes allows to manage complexity and flexible adjustment of behavior.

We intend to engineer CPSs with multi-modal software architectures to bridge the interval between the time zero-day vulnerabilities become known and the time updates become available. Modes can be switched to minimize the attack surface. This can be done automatically if a system detects threats by itself, or manually by an operator.

Early work on our SLR has provided a first impression about existing work on multiple modes and mode-switching. We want to adapt our SLR and take the scientific databases Wiley, Springer and Scopus into account. Next, we plan to use multiple modes in the context of security. For that purpose, we plan to build a multi-mode prototype and to demonstrate its ability to react to various threats and attacks.

Acknowledgement. This work has partially been supported by the LIT Secure and Correct Systems Lab funded by the State of Upper Austria.

References

1. Abeni, L., Buttazzo, G.: Hierarchical QoS management for time sensitive applications. In: Proceedings Seventh IEEE Real-Time Technology and Applications Symposium, pp. 63–72 (2001). https://doi.org/10.1109/RTTAS.2001.929866
2. Andersson, B.: Uniprocessor EDF scheduling with mode change. In: Baker, T.P., Bui, A., Tixeuil, S. (eds.) OPODIS 2008. LNCS, vol. 5401, pp. 572–577. Springer, Heidelberg (2008). https://doi.org/10.1007/978-3-540-92221-6_43
3. Bailey, C.: Hard real time operating system kernel: investigation of mode change, task 14 deliverable on estsec contract 9198/90/nl. sf, Technical report, British Aerospace Systems Ltd. (1993)
4. Block, A., Anderson, J.H., Devi, U.C.: Task reweighting under global scheduling on multiprocessors. Real-Time Syst. **39**(1), 123–167 (2008). https://doi.org/10.1007/s11241-007-9041-2
5. Borde, E., Haik, G., Pautet, L.: Mode-based reconfiguration of critical software component architectures. In: Automation Test in Europe Conference Exhibition 2009 Design, pp. 1160–1165 (2009). https://doi.org/10.1109/DATE.2009.5090838
6. Burns, A., Davis, R.I., Baruah, S., Bate, I.: Robust mixed-criticality systems. IEEE Trans. Comput. **67**(10), 1478–1491 (2018). https://doi.org/10.1109/TC.2018.2831227
7. Capota, E.A., Stangaciu, C.S., Micea, M.V., Curiac, D.I.: Towards mixed criticality task scheduling in cyber physical systems: challenges and perspectives. J. Syst. Softw. **156**, 204–216 (2019). https://doi.org/10.1016/j.jss.2019.06.099

8. Chen, T., Phan, L.T.X.: SafeMC: a system for the design and evaluation of mode-change protocols. In: 2018 IEEE Real-Time and Embedded Technology and Applications Symposium (RTAS), pp. 105–116 (2018). https://doi.org/10.1109/RTAS.2018.00021

9. Firesmith, D.: System resilience: what exactly is it? (2019). https://insights.sei.cmu.edu/sei_blog/2019/11/system-resilience-what-exactly-is-it.html

10. Hang, Y., Hansson, H.: Handling emergency mode switch for component-based systems. In: 2014 21st Asia-Pacific Software Engineering Conference, vol. 1, pp. 151–158 (2014). https://doi.org/10.1109/APSEC.2014.32

11. Hanninen, K., Maki-Turja, J., Nolin, M., Lindberg, M., Lundback, J., Lundback, K.L.: The Rubus component model for resource constrained real-time systems. In: 2008 International Symposium on Industrial Embedded Systems, pp. 177–183 (2008). https://doi.org/10.1109/SIES.2008.4577697

12. Hansson, H., AAkerholm, M., Crnkovic, I., Torngren, M.: SaveCCM - a component model for safety-critical real-time systems. In: Proceedings. 30th Euromicro Conference, 2004, ppD. 627–635 (2004). https://doi.org/10.1109/EURMIC.2004.1333431

13. Henzinger, T.A., Horowitz, B., Kirsch, C.M.: Giotto: a time-triggered language for embedded programming. In: Henzinger, T.A., Kirsch, C.M. (eds.) EMSOFT 2001. LNCS, vol. 2211, pp. 166–184. Springer, Heidelberg (2001). https://doi.org/10.1007/3-540-45449-7_12

14. Hirsch, D., Kramer, J., Magee, J., Uchitel, S.: Modes for software architectures. In: Gruhn, V., Oquendo, F. (eds.) EWSA 2006. LNCS, vol. 4344, pp. 113–126. Springer, Heidelberg (2006). https://doi.org/10.1007/11966104_9

15. Ke, X., Sierszecki, K., Angelov, C.: COMDES-II: a component-based framework for generative development of distributed real-time control systems. In: 13th IEEE International Conference on Embedded and Real-Time Computing Systems and Applications (RTCSA 2007), pp. 199–208 (2007). https://doi.org/10.1109/RTCSA.2007.29

16. Kitchenham, B., Charters, S.: Guidelines for performing systematic literature reviews in software engineering (version 2.3). Technical report, EBSE-2007-01, Keele University and Durham University (2007)

17. Maraninchi, F., Rémond, Y.: Mode-automata: about modes and states for reactive systems. In: Hankin, C. (ed.) ESOP 1998. LNCS, vol. 1381, pp. 185–199. Springer, Heidelberg (1998). https://doi.org/10.1007/BFb0053571

18. Maraninchi, F., Rémond, Y.: Mode-Automata: a new domain-specific construct for the development of safe critical systems. Sci. Comput. Program. **46**(3), 219–254 (2003). https://doi.org/10.1016/S0167-6423(02)00093-X

19. Martins, P., Burns, A.: On the meaning of modes in uniprocessor real-time systems. In: Proceedings of the 2008 ACM Symposium on Applied Computing, SAC 2008, pp. 324–325. Association for Computing Machinery (2008). https://doi.org/10.1145/1363686.1363770

20. McGraw, G.: Software security. IEEE Secur. Priv. **2**, 80–83 (2004)

21. Meumeu Yomsi, P., Nelis, V., Goossens, J.: Scheduling multi-mode real-time systems upon uniform multiprocessor platforms. In: 15th IEEE International Conference on Emerging Technologies and Factory Automation (ETFA 2010), pp. 1–8 (2010). https://doi.org/10.1109/ETFA.2010.5641275

22. Nelis, V., Andersson, B., Marinho, J., Petters, S.M.: Global-EDF scheduling of multimode real-time systems considering mode independent tasks. In: 2011 23rd Euromicro Conference on Real-Time Systems, pp. 205–214 (2011). https://doi.org/10.1109/ECRTS.2011.27

23. Nelis, V., Goossens, J., Andersson, B.: Two protocols for scheduling multi-mode real-time systems upon identical multiprocessor platforms. In: Proceedings - Euromicro Conference on Real-Time Systems, pp. 151–160 (2009). https://doi.org/10.1109/ECRTS.2009.27

24. van Ommering, R., van der Linden, F., Kramer, J., Magee, J.: The koala component model for consumer electronics software. Computer **33**(3), 78–85 (2000). https://doi.org/10.1109/2.825699

25. Pedro, P., Burns, A.: Schedulability analysis for mode changes in flexible real-time systems. In: Proceeding. 10th EUROMICRO Workshop on Real-Time Systems (Cat. No.98EX168), pp. 172–179 (1998). https://doi.org/10.1109/EMWRTS.1998.685082

26. Pedro, P.S.M.: Schedulability of mode changes in flexible real-time distributed systems. Ph.D. thesis, University of York, Department of Computer Science (1999)

27. Petticrew, M., Roberts, H.: Systematic Reviews in the Social Sciences: A Practical Guide, vol. 11. Wiley (2006). https://doi.org/10.1002/9780470754887

28. Phan, L.T., Lee, I.: Towards a compositional multi-modal framework for adaptive cyber-physical systems. In: in Proceedings of the 17th International Conference on Embedded and Real-Time Computing Systems and Applications, pp. 67–73. IEEE (2011). https://doi.org/10.1109/RTCSA.2011.82

29. Rao, A., Carreón, N., Lysecky, R., Rozenblit, J., Sametinger, J.: Resilient security of medical cyber-physical systems. In: Anderst-Kotsis, G., et al. (eds.) DEXA 2019. CCIS, vol. 1062, pp. 95–100. Springer, Cham (2019). https://doi.org/10.1007/978-3-030-27684-3_13

30. Rao, A., Rozenblit, J., Lysecky, R., Sametinger, J.: Trustworthy multi-modal framework for life-critical systems security. In: Proceedings of the Annual Simulation Symposium, ANSS 2018, pp. 1–9. Society for Computer Simulation International (2018)

31. Real, J.: Protocolos de cambio de modo para sistemas de tiempo real (mode change protocols for real time systems). Ph.D. thesis, Universitat Politècnica de València (2000). https://dialnet.unirioja.es/servlet/tesis?codigo=8892

32. Real, J., Crespo, A.: Mode change protocols for real-time systems: a survey and a new proposal. Real-Time Syst. **26**(2), 161–197 (2004). https://doi.org/10.1023/B:TIME.0000016129.97430.c6

33. Resmerita, S., Derler, P., Pree, W.: Timing Definition Language (TDL) Modeling in Ptolemy II. Technical report 21, Department of Computer Science, University of Salzburg (2020)

34. Sametinger, J., Steinwender, C.: Resilient context-aware medical device security. In: International Conference on Computational Science and Computational Intelligence, Symposium on Health Informatics and Medical Systems (CSCI-ISHI), pp. 1775–1778 (2017). https://doi.org/10.1109/CSCI.2017.310. http://americancse.org/events/csci2017/Symposiums/csci-ishi

35. Schoeberl, M.: Mission modes for safety critical Java. In: Obermaisser, R., Nah, Y., Puschner, P., Rammig, F.J. (eds.) SEUS 2007. LNCS, vol. 4761, pp. 105–113. Springer, Heidelberg (2007). https://doi.org/10.1007/978-3-540-75664-4_11

36. Sha, L., Goodenough, J.B.: Real-time scheduling theory and Ada. Computer **23**(4), 53–62 (1990). https://doi.org/10.1109/2.55469

37. Sha, L., Rajkumar, R., Lehoczky, J., Ramamritham, K.: Mode change protocols for priority-driven preemptive scheduling. Real-Time Syst. **1**(3), 243–264 (1989). https://doi.org/10.1007/BF00365439

38. Shih, C.S., Yang, C.M., Su, W.L., Tsung, P.K.: OSAMIC: online schedulability analysis of real-time mode change on heterogeneous multi-core platforms. In: Proceedings of the 2018 Conference on Research in Adaptive and Convergent Systems, RACS 2018, pp. 205–212. ACM (2018). https://doi.org/10.1145/3264746.3264755
39. Sundar, V.K., Easwaran, A.: A practical degradation model for mixed-criticality systems. In: 2019 IEEE 22nd International Symposium on Real-Time Distributed Computing (ISORC), pp. 171–180 (2019). https://doi.org/10.1109/ISORC.2019.00040
40. Søndergaard, H., Ravn, A.P., Thomsen, B., Schoeberl, M.: A practical approach to mode change in real-time systems. Technical report 08–001, Department of Computer Science, Aalborg University (2008)
41. Tindell, K.W., Burns, A., Wellings, A.J.: Mode changes in priority pre-emptively scheduled systems. In: Proceedings of the Real Time Systems Symposium, pp. 100–109 (1992)
42. Tindell, K., Alonso, A.: A very simple protocol for mode changes in priority pre-emptive systems. Technical report, Universidad Politécnica de Madrid (1996)
43. Tiwari, A., et al.: Safety envelope for security. In: Proceedings of the 3rd International Conference on High Confidence Networked Systems, HiCoNS 2014, pp. 85–94. Association for Computing Machinery (2014). https://doi.org/10.1145/2566468.2566483

Exploiting MQTT-SN for Distributed Reflection Denial-of-Service Attacks

Hannes Sochor$^{(\boxtimes)}$ (ID), Flavio Ferrarotti (ID), and Rudolf Ramler (ID)

Software Competence Center Hagenberg GmbH, Softwarepark 21,
4232 Hagenberg, Austria
{hannes.sochor,flavio.ferrarotti,rudolf.ramler}@scch.at

Abstract. Distributed Denial-of-Service attacks are a dramatically increasing threat to Internet-based services and connected devices. In the form of reflection attacks they are abusing other systems to perform the actual attack, often with an additional amplification factor. In this work we describe a reflection attack exploiting the industrial Message Queuing Telemetry Transport for Sensor Networks (MQTT-SN) protocol, which theoretically allows to achieve an unlimited amplification rate. This poses a significant risk not only for the organizations which are running a MQTT-SN broker but also for possible targets of such DRDoS attacks. Countermeasures are limited as the underlying weakness is rooted in the specification of MQTT-SN itself.

Keywords: Security · Reflection attack · DDoS · DRDoS · MQTT-SN

1 Introduction

The number and intensity of denial-of-service (DoS) attacks has dramatically increased and it poses a major threat to the stability and reliability of Internet-based services and connected devices [1]. One of the largest DoS attacks ever observed was executed in February 2018 against the popular source code hosting platform Github, causing incoming malicious traffic of about 1.3 terabytes per second at its peak [3]. Another large-scale attack targeted the DNS provider Dyn in October 2016, resulting in outages of Netflix, Visa, Amazon and other major sites. The attack was launched using the botnet Mirai [2] consisting primarily of Internet of Things (IoT) devices.

In both examples, other systems have been exploited to simultaneously perform the actual attack in a distributed fashion. In this paper, we describe a related approach for attacks based on a weakness in the MQTT-SN protocol. We found that MQTT-SN is particularly susceptible to misuse. Hence, in contrast to other attacks, the exploit does not require tampering of the infrastructure. In fact, the design of the MQTT-SN protocol and its lack of common security mechanisms are enough to make any MQTT-SN broker a potential target that can be abused for attacking other systems.

© Springer Nature Switzerland AG 2020
G. Kotsis et al. (Eds.): DEXA 2020 Workshops, CCIS 1285, pp. 74–81, 2020.
https://doi.org/10.1007/978-3-030-59028-4_7

The overall goal of this paper is to increase the awareness about these so-called reflection attacks and the critical vulnerability of IoT-based systems. The paper is structured as follows: Sect. 2 provides an overview of the relevant background on DoS attacks and the MQTT-SN protocol. In Sect. 3 the attack and our prototypical implementation in a lab setting are described. Section 4 discusses possible countermeasures and protection mechanisms. Finally, Sect. 5 summarizes the paper and outlines future work.

2   Background

In this section we briefly review the relevant background. We start by introducing the concept of a denial-of-service (DoS) attack, followed by a description of the reflective and distributed variants of this attack. Finally we provide a short explanation of the MQTT protocol and its cousin MQTT-SN, emphasising the key security weakness of the latter which we exploit in this paper.

2.1   DoS Attack

The intention of a DoS attack is to prevent the target system from fulfilling its tasks. The target may be any component of the system such as applications, servers or other network components. A DoS attack can be achieved through different means. For instance, it can be achieved by crashing an application with malicious inputs, by consuming all available resources of a server, or by overloading the network with traffic so that the communication is disrupted.

2.2   Reflection Attack

A reflection attack is a kind of DoS attack, which attempts to consume all the bandwidth of the target system. Therefore, an attacker sends some valid requests to a server. When sending the request the attacker spoofs her own IP-address to that of her target. As a result the server sends its response to the initial request to the attacker's target. If the response is larger than the initial request, the attacker has achieved an *amplification* of the sent data. An illustrative example of such an attack is depicted in Fig. 1. In a real world scenario such as the one discussed in [4], the attacker is able to use as little as 0.02% of the bandwidth received by her victim. An example from the IoT environment is the reflection attack on the Constrained Application Protocol (CoAP), which is described in [5]. In addition to a possible amplification, the attacker efficiently hides her own IP-address, so that the target is not able to identify the source of the attack.

2.3   Distributed Denial-of-Service (DDoS) Attack

A DDoS attack is an attack that involves several distributed components. A classical example is the use of a botnet (see e.g. [2]) to simultaneously send requests to a server to overload its capacities.

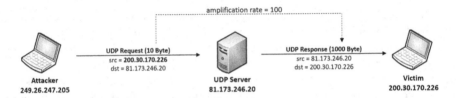

Fig. 1. The attacker sends a request to a server and spoofs her IP-address to that of the victim. The server then sends its response to the target of the attack. If the response is larger than the request, an amplification of the sent data has been achieved.

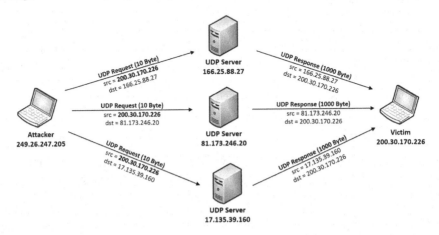

Fig. 2. An attacker performs a DRDoS attack by executing multiple reflection attacks simultaneously.

2.4 Distributed Reflective Denial-of-Service (DRDoS) Attack

When performing a simple reflection attack, one is limited by the network bandwidth of the used server. An attacker only uses a small fraction of her own network bandwidth. To use her full capacity the attacker may perform a reflection attack on more than one server simultaneously, which is called a DRDoS attack. This type of attack is depicted in Fig. 2.

2.5 MQTT and MQTT-SN

The MQTT protocol implements Machine to Machine (M2M) communication by enabling clients to exchange data via a central MQTT broker. A client may publish a message to the broker within the context of a set topic. Another client may subscribe to this topic and receive all published values.

When sending requests to the broker, the client decides on a quality of service (QoS) level for the communication. The QoS level defines which responses are expected and which handshakes are used. For instance, the broker accepts a PUBLISH message with QoS 0 without sending a response. With QoS 1 the response is a PUBACK. When using the highest level QoS 2, the client performs a four-way handshake with the broker.

MQTT for Sensor Networks [6] (MQTT-SN) is a variation of the MQTT protocol aimed at embedded devices. While the standard MQTT protocol relies in TCP to transmit packets, the MQTT-SN protocol uses UDP. The fact that the client can decide which QoS she wants to use (no responses, no handshake), coupled with the fact that MQTT-SN uses UDP to transmit packets, means that it is relatively easy to perform IP-address spoofing on MQTT-SN. This opens up the MQTT-SN protocol to reflection attacks.

3   DRDoS Attack Using MQTT-SN

In this section we describe a DRDoS attack which, using IP-address spoofing, takes advantage of the publish-subscribe functionality of MQTT-SN. This functionality enables a client to subscribe to a topic in order to receive all messages submitted to that particular topic by other clients. For this attack to proceed, we need a publicly available MQTT-SN broker. An overview of the attack performed on a single MQTT-SN broker is shown in Fig. 3. The concrete steps of this attack are as follows:

1. We prepare the attack by connecting some clients to the MQTT-SN broker and spoofing our source IP to the target IP address. From the point of view of the MQTT-SN broker, these clients are connected from the IP-address of our target.
2. Next we choose a topic and subscribe all the clients that we connected in the first step to that topic. Same as the connection, the subscription is done with the "spoofed" IP-address of our target.
3. Finally, we publish an arbitrary message to the topic chosen in the previous step. As a result the MQTT-SN broker publishes this message to all connected and subscribed clients. Since for the broker all these clients are connected from the IP address of our target, the targeted IP will get as many PUBLISH messages as clients we have subscribed in the previous step. We can repeat this last step, which allows us to easily consume all the bandwidth of the targeted IP.

If we execute the steps above on multiple MQTT-SN broker simultaneously, we can perform a full scale DRDoS attack with any desired amplification rate. The only limit to this amplification rate is given by the number of simultaneous clients that we can connect to the broker, which a priori is unbounded. This number may however vary greatly for different implementations of the MQTT-SN broker and available resources.

So far, we have implemented the attack in a virtual test environment using the Really Small Message Broker[1] (RSMB) MQTT-SN broker implementation. We tested the attack with a maximum of 10, 000 clients connected to each of our

---

[1] https://github.com/eclipse/mosquitto.rsmb.

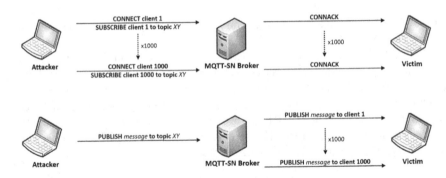

Fig. 3. We prepare the attack by connecting some clients to the MQTT-SN broker and then subscribing all those clients to a same topic (in both cases spoofing our source IP to the targeted IP address). We can then perform the reflection attack by publishing on that topic.

RSMB broker instances. This resulted in an amplification rate of almost 10,000, even when discounting the messages used for preparing the attack and the UDP packets lost or discarded on the way.

It is interesting to note that the amplification rate which we observed in our test environment is much higher than the average amplification rates attained for other protocols. In fact, the amplification rates reported in [4] are between 10 and 50, with the only exception of the memcached protocol for which the peak has been at 50,000.It is interesting to note that the amplification rate which we observed in our test environment is much higher than the average amplification rates attained for other protocols. In fact, the amplification rates reported in [4] are between 10 and 50, with the only exception of the memcached protocol for which the peak has been at 50,000.

4   Countermeasures and Protection Mechanisms

The weakness that enables the described reflection attack is rooted in the MQTT-SN protocol specification itself. As such all MQTT-SN broker implementations are affected as long as they are compliant with the specification. Thus additional external measures are required to protect a MQTT-SN brokers from this threat.

Enforcing additional handshakes in the protocol (e.g., by disallowing QoS levels 0 and 1) merely increases the effort required to perform the attack, but it does not prevent it completely. QoS level 2 guarantees that each message is received exactly once. This is ensured by means of a four-way handshake (PUBLISH – PUBREC – PUBREL – PUBCOMP) between the sender and the receiver. However, this does not suffice to completely prevent the attack from happening. It is still possible to send a spoofed PUBREL with a slight delay after the initial PUBLISH to deceive the broker.

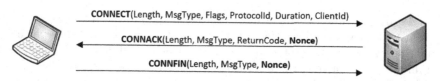

Fig. 4. Extending the MQTT-SN connection protocol from a two-way handshake into a three-way handshake with a random nonce. Required additions to the current protocol are highlighted in red. (Color figure online)

A better solution would be for the broker to enforce a three-way handshake for client connections. For instance, the broker could respond to a connection attempt with a randomly generated nonce (as commonly used in authentication protocols to ensure that old communications cannot be reused in replay attacks). The once could be sent to the client as part of the CONNACK package. Only if the client answers with a message (e.g. a CONNFIN) which includes the correct nonce, then the connection attempt would be successful. This would required changes to the MQTT-SN protocol itself as depicted in Fig. 4. In this schema, if an attacker attempts to spoof the IP address of a client, the CONNACK will be sent to the spoofed IP, rendering the connection attempt by the legitimate client unsuccessful. In theory this would prevent an attacker from connecting rouge clients, provided the messages are properly encrypted. If the messages are not encrypted, then an attacker could clearly obtain the nonce by sniffing on the network connection. An advantage of this approach is that only minimal changes to the protocol would be required and that those changes would only affect the part that deals with connections.

Of course, changes to the MQTT-SN standard involve a long and uncertain process. We should then think on possible workarounds that do not involve changes to the protocol and can effectively prevent DRDoS attacks. Some alternatives we can think of are the following:

- **Compulsory authentication.** If the broker implementation supports authentication, we can enforce it for all connections. This would prevent an attacker to connect the necessary clients to carry on the DRDoS attack, unless he could somehow get hold of the credentials.
- **White-listed clients.** If we know beforehand the IP addresses of authorized clients (or at least their IP range), then the server can simply reject connections from unauthorized IP addresses. This prevents the connection of rouge clients from IP addresses that are not white-listed, making the implementation of DRDoS attacks more difficult.
- **Bounded active connections.** Some broker implementations can impose limits to the number of simultaneous active connections. Although this does not prevent an attacker from connecting rouge clients, it can significantly restrict the potential for producing harm.

If we are presented with a situation in which we have no influence over the implementation and configurations of the broker, that is, a situation in which we cannot apply any of the protection mechanisms described above, then we need to think on different mitigation measures similar to those used for the related problem of mitigation of DDOS attacks (see for instance [7]). A general approach in this sense could be based on identifying suspected DDOS traffic, sending this traffic through a high capacity network were the malicious messages can be safely filtered. In our case this should be rather straightforward since such traffic would consist of nearly identical MQTT publish messages sent via UDP.

5   Summary and Future Work

We have described in detail how a successful distributed DoS reflection attack can be carried out over any system using MQTT-SN as communication protocol. It is a noteworthy fact that by means of this attack, a single small message can be multiplied an unbounded number of times, easily clogging up the victim's network. What makes this attack rare in comparison to other reflection attacks [4] is the (theoretically) unlimited amplification that can be achieved.

The described attack highlights the inherent weaknesses in the MQTT-SN protocol, as MQTT-SN lacks support for security mechanisms like simple authentication. Consequently, in order to improve security and to avoid attacks, the weaknesses in the protocol itself have to be targeted. Furthermore, although MQTT-SN is a common and widely used protocol in IoT environments, it is only one example of a wide range of protocol used for machine-to-machine communication. Similar weaknesses may also be found in related protocols.

As future work, we plan to: (a) Investigate and quantify the actual exposure of publicly available sensor networks that rely on the MQTT-SN protocol and are affected by this inherent vulnerability; (b) determine which other publish/subscribe protocols suffer from similar weaknesses and are prone to related exploits; and (c) propose general viable solutions to protect sensor networks that rely on publish/subscribe protocols from this kind of vulnerabilities.

Acknowledgement. The research reported in this paper has been supported by the Austrian Research Promotion Agency (FFG) within the ICT of the Future grants program (grant #863129, project IoT4CPS) and the COMET - Competence Centers for Excellent Technologies Programme (grant #865891, SCCH) funded by the Federal Ministry for Climate Action, Environment, Energy, Mobility, Innovation and Technology (BMK), the Federal Ministry for Digital and Economic Affairs (BMDW), and the Province of Upper Austria.

References

1. Jonker, M., King, A., Krupp, J., Rossow, C., Sperotto, A., Dainotti, A.: Millions of targets under attack: a macroscopic characterization of the dos ecosystem. In: Proceedings of the 2017 Internet Measurement Conference, IMC 2017, pp. 100–113. ACM, New York (2017)

2. Kolias, C., Kambourakis, G., Stavrou, A., Voas, J.: DDoS in the IoT: Mirai and other botnets. Computer **50**(07), 80–84 (2017)
3. Majkowski, M.: Memcrashed - major amplification attacks from UDP port 11211 (2018). https://blog.cloudflare.com/memcrashed-major-amplification-attacks-from-port-11211/. Accessed 14 Apr 2020
4. Rossow, C.: Amplification hell: revisiting network protocols for DDoS abuse. In: In: Proceedings of the 2014 Network and Distributed System Security Symposium, NDSS (2014)
5. Shelby, Z., Hartke, K., Bormann, C.: The constrained application protocol (COAP). RFC 7252, RFC Editor, June 2014. http://www.rfc-editor.org/rfc/rfc7252.txt, http://www.rfc-editor.org/rfc/rfc7252.txt
6. Stanford-Clark, A., Truong, H.L.: MQTT For Sensor Networks (MQTT-SN). IBM. http://mqtt.org/new/wp-content/uploads/2009/06/MQTT-SN_spec_v1.2.pdf
7. Yan, Q., Huang, W., Luo, X., Gong, Q., Yu, F.R.: A multi-level DDoS mitigation framework for the industrial Internet of Things. IEEE Commun. Mag. **56**(2), 30–36 (2018)

Machine Learning and Knowledge Graphs

Exploring the Influence of Data Aggregation in Parking Prediction

Shereen Elsayed, Daniela Thyssens$^{(\boxtimes)}$, Shabanaz Chamurally, Arslan Tariq, and Hadi Samer Jomaa

University of Hildesheim, Universitätspl. 1, 31141 Hildesheim, Germany
{elsayed,thyssens,chamural,tariqh,jomaah}@uni-hildesheim.de
https://www.uni-hildesheim.de/en/

Abstract. Parking occupancy is influenced by many external factors that make the availability prediction task difficult. We want to investigate how information from different data sources, such as events, weather and geographical entities interrelate in affecting parking prediction and thereby form a knowledge graph for the parking prediction problem.

In this paper, we try to tackle this problem by answering the following questions; What is the effect of the external features on different models? Is there a correlation between the amount of historical training data and external features? These questions are evaluated by applying three well-known time series forecasting models; long short term memory, convolutional neural network and multilayer perceptron. Additionally we introduce gradient boosted regression trees with handcrafted features. Experimental results on two real-world datasets showed that external features have a significant effect throughout the experiments and that the extent of the effectiveness varies across training histories and tested models. The findings show that the models are able to outperform recent work in the parking prediction literature. Furthermore, a comparison of the feature-engineered gradient boosted decision trees to other potential models has shown its advantage in the field of time series forecasting.

Keywords: Time series forecasting · Parking occupancy · Machine learning

1 Introduction

The search for available parking spaces in central business district areas generates numerous disadvantages with regards to the traffic efficiency and ecological externalities, such as city air-pollution and parking search traffic. Moreover, various sources attest that over one-third of the upcoming congestions originate from searching for free parking spaces [10,13] constituting a major inefficiency in terms of time costs for citizens. Providing timely availability information concerning

---

S. Elsayed and D. Thyssens—Both authors contributed equally to this research.

G. Kotsis et al. (Eds.): DEXA 2020 Workshops, CCIS 1285, pp. 85–95, 2020.
https://doi.org/10.1007/978-3-030-59028-4_8

city car parks can mitigate both, ecological and time-efficiency drawbacks by suggesting vacant parking spaces to drivers before they commence their journey.

With regards to these motivational grounds, a recently growing literature applies deep learning (DL) approaches to the time series forecasting problem, achieving noticeable gains concerning the overall forecasting performance [3,14,15]. A number of these approaches integrate external context information [8,14,15], but while these approaches emphasize the nature, in which they are introduced to the given model, it remains largely unstudied how the effectiveness of these features varies across different models and different amounts of historical data.

In this work, we therefore investigate the effect of external features on the prediction of parking occupancy based on different amounts of training data considering four models: long short term memory (LSTM), convolutional neural network (CNN), and multilayer perceptron (MLP) and gradient boosted regression trees (XGboost). The aim is to identify differences in the selected models under varying data availability circumstances and accentuate the effectiveness of external context features in this regard. The findings support the effectiveness of external features in parking occupancy prediction and underline that the extent and direction of the effect greatly depend on the model and amount of training data selected. To this end, XGBoost outperforms other models in the long run given the additional features. Our results are substantiated by a comparative analysis regarding the work by Camero et al. [3], where the selected models in the present work outperform those in [3]. Moreover, we compare the feature-engineered XGBoost to DL models in other time series forecasting fields and reveal its competitiveness to thoroughly elaborated DL frameworks, such as the deep air quality forecasting framework (DAQFF) [6].

The remainder of the paper is structured as follows: Sect. 2 reviews the existing approaches in the field of parking occupancy prediction. Section 3 introduces the followed methodology. Section 4 elaborates on the experiments, discusses some preliminary feature analysis and documents the results, whereas Sect. 5 concludes.

2   Related Work

Being part of the family of time series forecasting problems, parking occupancy prediction constitutes an extensive body of research. The designed approaches range from regression-based methods to neural network (NN) models. This section comprises an overview of the recent developments regarding parking availability prediction with a particular focus on machine learning models.

The majority of the recent works in the area of parking occupancy prediction present distinct NN-based learning approaches [3,7,11,12], for example concerning real-time occupancy prediction; Rong et al. [11] employ a DL approach (Du-Parking) that models temporal closeness, period and general influence to estimate the final parking availability of parking lots in nine Chinese cities. The same type of methodological approaches can be found in the traffic prediction field; numerous works center NNs in their approach, capturing spatial-temporal correlations by integrating these models into end-to-end architectures to process heterogeneous time series, such as weather and events time series, in a unified framework [9,14,16]. This way of processing heterogeneous time series data into single NN models has inspired the architecture of the proposed LSTM model structure in Sect. 3.

Nevertheless, there exist other ways to proceed; Probabilistic approaches in the parking occupancy field [1,2,8] concentrate mainly on modeling continuous-time Markov models, where Beheshti et al. [1] focus on a hybridization of a Markov chain Monte Carlo model with an agent-based model to predict parking and traffic fluctuations.

Closer to our work, Zheng et al. [17] analyze three non-parametric models (regression tree, NN, support vector regression (SVR)) based on three feature sets, concluding that given its small computational costs, regression trees performs best for predicting parking occupancy. Chen [4] explores the effects of aggregation on parking occupancy prediction by evaluating different models (ARIMA, SVR, NN) on data from San Francisco. Camero et al. [3] focus on meta-heuristic approaches regarding the optimization of Recurrent NNs to predict parking occupancy. What distinguishes the present work from these studies is, on one hand, the selection of models, which are extensions or modified versions of those used in the studies mentioned above. On the other hand, we additionally incorporate external context information and enlarge the experimental scope by employing scalability experiments.

3   Methodology

The goal is to study and evaluate the influence of the external features on parking occupancy using different models. This section is dedicated to the formulation of the research problem and to the elaboration of the used models.

3.1   Problem Formulation

The occupancy prediction problem intends to predict the occupancy at time $t + h$ given data from the time interval $t - l$ to t, where h is the forecasting horizon and l is the history. The input instances are considered as a vector $(y_{t-l,...,t}, f_{t-l,...,t})$, where $y_{t-l,...,t} \in \mathbf{R}^l$ is the historical occupancy data and $f_{t-l,...,t} \in \mathbf{R}^{l \times J}$ with feature dimension J, denotes the external features to predict a vector $(y_{t+1}, \ldots, y_{t+h})$. We define the loss function over N instances as follows and optimized it using Adam optimizer.

$$\mathcal{L}(\theta) = \sum_{i=1}^{N} \sum_{k=1}^{h} (y_{t+k} - \hat{y}_{t+k})^2 + \lambda \|\theta\|_2 \tag{1}$$

3.2 LSTM

The LSTM model as shown in Fig. 1 consists of three main components, each handling different types (time, static and event) of input features. The input-data is cut into windows consisting of two hours in order to predict one hour. Static features are encoded by two fully connected layers before being introduced into a single LSTM layer of 128 neurons and ReLu activation function together with the time features [9,15]. Event features are taken into a separate one-layer LSTM with 128 neurons and ReLu activation function, before both outputs, are combined and fed into a final fully connected layer for the multi-output prediction.

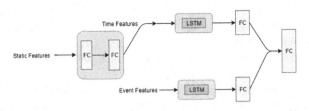

Fig. 1. LSTM model.

3.3 CNN

As for the LSTM model, the input data is divided into time windows. As shown in Fig. 2), all types of features are simultaneously fed into two 1D-CNN layers. A fully connected layer is implemented on the flattened output to get the final multi-output prediction. We investigated in different network structures concerning the CNN; a simple one, as described above, proved to work very well in comparison to more extensive structures.

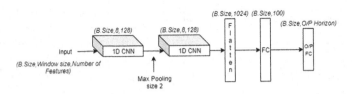

Fig. 2. CNN model.

3.4   XGBoost

The data is introduced, in a distinct windowed fashion producing continuous multi-output predictions to allow direct comparability with the predictions obtained from LSTM and CNN. Figure 3 shows how the time windows are transformed into 2D instances, in which the target feature, occupancy (red blocks), is concatenated with the last time step of the explanatory features (grey block). We conducted empirical evaluations on how many time-steps, regarding the explanatory features should be included in the 2D representation of the instance and concluded that the last time step only provides the best results concerning the prediction- and computational-performance. These 2D instances are then taken as input for the multi-output XGBoost to get the prediction corresponding to each of the input instances.

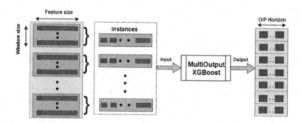

Fig. 3. Feature engineered XGBoost model. (Color figure online)

3.5   MLP

Finally, as a basic baseline model, we employ an MLP that consists of a simple fully connected layer, where the input is again a sequence of 2D instances similar to XGBoost. In our implementation, the fully connected layer comprises 100 neurons followed by a second layer that is composed of 50 neurons and the activation function ReLu. Lastly, another fully connected layer for the multi-output prediction is instantiated.

4   Experiments

This section discusses the setup, datasets and extracted features for the main experiments, before the results are demonstrated and analyzed along the following research questions:

RQ1: What is the effect of external features on different models?

RQ2: Is there a correlation between the amount of historical training data and external features?

RQ3: Is the performance of the selected models comparable to that in other recent parking prediction works?

RQ4: How competitive is XGBoost against DL models?

4.1   Evaluation Protocol

We cut the data time-consistently into training, validation and testing (80, 10, 10% respectively). Our objective is to minimize the RMSE between the actual occupancy Y and the predicted occupancy Y'. Each RMSE value is validated by five identical runs for each experiment and hyperparameters were tuned for each model individually in an empirical fashion. The structural model parameters in the model descriptions above were giving the best results respectively.

$$RMSE = \sqrt{\frac{1}{N}(Y - Y')^2} \tag{2}$$

4.2   Datasets

The subsequent, publicly available datasets are used to evaluate our research proposition.

Banes Dataset: The Banes Historic Car Park Occupancy[1] is provided by the Bath and North East Somerset Council. It consists in total of 2.517.509 records, of which 1.256.976 were used, from eight different off-street car parks located in Bath, UK. The time series is updated every five minutes. For the first six parking locations we used the data collected for the years 2017 and 2018, otherwise, data from 2015 and 2016 was used due to incompleteness. Issues were discovered regarding temporary sensor malfunctions, reserved parking spots and similar failures. These problems were addressed as follows: The percentage values beyond 100% were adjusted to 100%, the absolute value of negative percentages was taken, out-of-date data and duplicate readings are discarded.

Birmingham Dataset: The UCI repository dataset "Parking in Birmingham"[2] comprises in total 35717 records of occupancy rates of 30 car parks in Birmingham, UK. The occupancy values were updated every 30 min during peak hours only, that is, from 7:00 to 16:30. We neglected the data concerning car park 8 and 21 as the majority of the data was missing.

4.3   Feature Extraction

Since a major part of this work involved the extraction and incorporation of external context features that were not included in the original datasets, we dedicate this section mainly to the illustration and elaboration of those gathered features.

---

[1] https://data.bathhacked.org/Transport/BANES-Live-Car-Park-Occupancy/u3w2-9yme.
[2] https://data.birmingham.gov.uk/dataset/birmingham-parking.

Temporal Features. Temporal features refer to the features that can be retrieved from the timestamp, at which the sensors recorded a particular parking occupancy. Additional to the original features contained in the dataset (timestamp, occupancy, capacity, etc.), we extracted the Year, Month, Day, Hour, Minute, "Is Weekend" and "Day of Week" as independent explanatory features. The feature "Is Weekend indicates if the occupancy rate was recorded during the weekends or working days and, "Day Of Week" indicates the weekday, encoded as a number ranging from 0–6.

Spatial Features. To include the influence of several spatial and geographical circumstances, we considered several physical points of interest that surround a particular parking location (number of schools, burial grounds, general practitioners and parks), which were retrieved using the Places API[3] provided by Google Inc. Since these features are deterministic per location and don't change over time, we consider them as *static*.

Meteorological Features. Meteorological changes influence the parking availability on multiple levels; e.g. when it rains, chances are that the demand for parking spaces in the city increases as citizens prefer to drive to the city, whereas it might also be the case that on rainy days, people prefer to stay at home while they can, thereby reducing parking occupancy. We included several meteorological features, such as temperature and humidity as well as weather conditions (e.g. sunny, windy, rainy, etc.).

Event Features. We considered for example Football and Rugby games by including a binary variable indicating whether such an event took place on the given day or not. The information was retrieved manually by researching the match fixtures for a given year on the web[4]. Figure 4 shows that events can also have adversarial effects on parking occupancy when the event location is not in the range of the parking locations considered, whereas Fig. 5 displays the feature importance across all features for the case of the Birmingham dataset.

4.4   Features Effect and Scalability Experiments (RQ1 and RQ2)

The results of the scalability experiments in Fig. 6 display an overall downward trend of the RMSE values considering more training history across models. XGBoost continuously benefits from both, more historical data and included features, even though it cannot make use of additional features for as little as one day of data. The opposite behavior can be observed for LSTM and CNN; the models benefit from the additional features regarding small training data sizes, but the feature effectiveness declines gradually with more training data, concluding that LSTM is mainly affected by history for prediction.

---

[3] https://cloud.google.com/maps-platform.
[4] http://bath.co.uk/event/bath-city-fc.

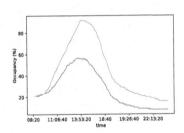

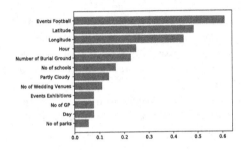

Fig. 4. Event day vs. no event day   **Fig. 5.** Feature importance (XGBoost)

The two main findings regarding the Banes dataset are emphasized by the results from the Birmingham dataset shown in Fig. 7. For MLP and CNN there is no common trend in the two datasets, but for the Birmingham data, MLP acts the same way as XGBoost. Hence, XGBoost is the model that performs best in the long run given the additional features, whereas LSTM does not increasingly benefit from the information in the features when extending the training history.

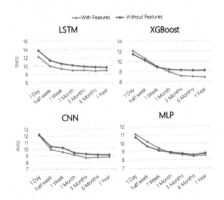

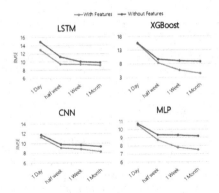

Fig. 6. Banes data results (forecasting: 1 h (12 instances), window size of 2 h)   **Fig. 7.** Birm. Data results (forecasting: 1 h (2 instances), window size: 2 h)

4.5 Comparison to Other Parking Prediction Models (RQ3)

The study we are comparing our work to, considers training each parking location in the Birmingham dataset separately and thus, despite it being computationally expensive, we followed the same evaluation scheme for the purpose of comparability.

Table 1 compares our configured models (in bold) to the employed RNN and the associated baselines (P, F, KM, KP, SP, TS) in [3]. On average, all the models considered in the present paper, except LSTM, outperform the ones in [3], where CNN performs best followed by XGBoost (XGB). Since separate models are trained for each parking location, this generally accounts for less

Table 1. Average, maximum and minimum mean average errors for predicting normalized occupancy rates across the parking locations on Birm. Data (forecasting horizon: 1 day (18 instances), window size: 3.5 h)

Models	P	F	KM	KP	SP	TS	RNN	LSTM	CNN	MLP	XGB
Mean	0.067	0.079	0.102	0.101	0.073	0.067	0.079	0.068	**0.019**	0.059	0.054
Max	0.132	0.148	0.177	0.179	0.139	0.129	0.137	0.077	0.025	0.104	0.104
Min	0.016	0.024	0.048	0.049	0.025	0.023	0.033	0.056	0.016	0.021	0.025

training data per model, in this way Table 1 supports the results from Fig. 7 in that LSTM does not perform optimally given limited training data, whereas CNN and MLP perform already fairly well. These results also support the notion of the effectiveness of feature aggregation on the Birmingham dataset.

4.6 Competitiveness of XGBoost in Multivariate Time Series Forecasting (RQ4)

We employ the feature-engineered XGBoost model on the problem of air quality prediction and compare it to DL-based models using the publicly available UCI repository dataset PM2.5[5]. Table 1 and Fig. 8 highlight the performance and competitiveness of XGBoost for time series forecasting. Both, the DAQFF [6] and the Time Series Sequence-to-sequence Deep learning Framework (TSDLF) [5] consists of complex forecasting frameworks, in which the main architectural focus is laid on LSTMs. Regarding Table 2, the columns are dedicated to different forecasting horizons, where a training window of 6 time-steps is used to forecast 1, 3 and 6 time-steps respectively. XGBoost outperforms the given baselines, as

Table 2. The table shows the RMSE of [5] vs XGBoost given different forecasting horizons and a lookup size 6, (w, h).

Models	(6,1)	(6,3)	(6,6)
SVR-POLY	0.068	0.116	0.098
SVR-RBF	0.045	0.055	0.067
SVR-LINEAR	0.034	0.057	0.072
LSTM	0.051	0.048	0.058
GRU	0.042	0.063	0.178
RNN	0.099	0.098	0.097
TSDLF	0.031	0.040	0.049
XGBoost	**0.022**	**0.033**	**0.044**

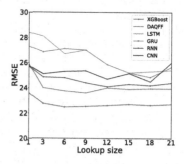

Fig. 8. Results of [6] vs XGBoost

well as the TSDLF in all forecasting horizons. Figure 8 shows that XGBoost outperforms the DAQFF for all indicated training windows (lookup sizes) by far.

5   Conclusion

Apart from constituting that the aggregated features generally affect parking occupancy prediction, the results of the experiments (on both datasets) above, put forth two notions regarding the extent and direction of this effect concerning the different models; For one thing, we see differences in the behavior of the models regarding the inclusion of additional features; here XGBoost greatly benefits not only from more past information but also from the external context features. On the other hand, we observe that the overall performance of LSTM remains behind expectations, especially in the long run when comparing it directly to XGBoost.

References

1. Beheshti, R., Sukthankar, G.: A hybrid modeling approach for parking and traffic prediction in urban simulations. AI Soc. **30**(3), 333–344 (2014). https://doi.org/10.1007/s00146-013-0530-7
2. Caliskan, M., Barthels, A., Scheuermann, B., Mauve, M.: Predicting parking lot occupancy in vehicular ad hoc networks. In: 2007 IEEE 65th Vehicular Technology Conference-VTC2007-Spring, pp. 277–281. IEEE (2007)
3. Camero, A., Toutouh, J., Stolfi, D.H., Alba, E.: Evolutionary deep learning for car park occupancy prediction in smart cities. In: Battiti, R., Brunato, M., Kotsireas, I., Pardalos, P.M. (eds.) LION 12 2018. LNCS, vol. 11353, pp. 386–401. Springer, Cham (2019). https://doi.org/10.1007/978-3-030-05348-2_32
4. Chen, X.: Parking occupancy prediction and pattern analysis. Technical report CS229-2014, Department of Computer Science, Stanford University, Stanford, CA, USA (2014)
5. Du, S., Li, T., Horng, S.J.: Time series forecasting using sequence-to-sequence deep learning framework. In: 2018 9th International Symposium on Parallel Architectures, Algorithms and Programming (PAAP), pp. 171–176. IEEE (2018)
6. Du, S., Li, T., Yang, Y., Horng, S.: Deep air quality forecasting using hybrid deep learning framework. IEEE Trans. Knowl. Data Eng. (01), 1–1, 5555 (2019). https://doi.org/10.1109/TKDE.2019.2954510
7. Ji, Y., Tang, D., Blythe, P., Guo, W., Wang, W.: Short-term forecasting of available parking space using wavelet neural network model. IET Intell. Transp. Syst. **9**(2), 202–209 (2014)
8. Klappenecker, A., Lee, H., Welch, J.L.: Finding available parking spaces made easy. Ad Hoc Netw. **12**, 243–249 (2014)
9. Laptev, N., Yosinski, J., Li, L.E., Smyl, S.: Time-series extreme event forecasting with neural networks at uber. In: International Conference on Machine Learning, vol. 34, pp. 1–5 (2017)
10. Ma, J., Clausing, E., Liu, Y.: Smart on-street parking system to predict parking occupancy and provide a routing strategy using cloud-based analytics. Technical report, SAE Technical Paper (2017)

11. Rong, Y., Xu, Z., Yan, R., Ma, X.: Du-Parking: spatio-temporal big data tells you realtime parking availability. In: Proceedings of the 24th ACM SIGKDD International Conference on Knowledge Discovery & Data Mining, pp. 646–654 (2018)
12. Shao, W., Zhang, Yu., Guo, B., Qin, K., Chan, J., Salim, F.D.: Parking availability prediction with long short term memory model. In: Li, S. (ed.) GPC 2018. LNCS, vol. 11204, pp. 124–137. Springer, Cham (2019). https://doi.org/10.1007/978-3-030-15093-8_9
13. Tilahun, S.L., Di Marzo Serugendo, G.: Cooperative multiagent system for parking availability prediction based on time varying dynamic Markov chains. J. Adv. Transp. **2017** (2017)
14. Yao, H., Tang, X., Wei, H., Zheng, G., Li, Z.: Revisiting spatial-temporal similarity: a deep learning framework for traffic prediction. In: Proceedings of the AAAI Conference on Artificial Intelligence, vol. 33, pp. 5668–5675 (2019)
15. Yao, H., et al.: Deep multi-view spatial-temporal network for taxi demand prediction. In: Thirty-Second AAAI Conference on Artificial Intelligence (2018)
16. Zhang, J., Zheng, Y., Qi, D.: Deep spatio-temporal residual networks for citywide crowd flows prediction. In: Thirty-First AAAI Conference on Artificial Intelligence (2017)
17. Zheng, Y., Rajasegarar, S., Leckie, C.: Parking availability prediction for sensor-enabled car parks in smart cities. In: 2015 IEEE Tenth International Conference on Intelligent Sensors, Sensor Networks and Information Processing (ISSNIP), pp. 1–6. IEEE (2015)

Building Knowledge Graph in Spark Without SPARQL

Alex Romanova[(⊠)]

Melenar, LLC, McLean, VA 22101, USA
sparkling.dataocean@gmail.com

Abstract. Knowledge graphs, powerful assets for enhancing search and various data integration, are being essential in both academia and industry. In this paper we will demonstrate that knowledge graph abilities are much wider than search and data integration. We will do it in a twofold manner: 1) we will show how to build knowledge graph in Spark instead of using SPARQL language and how to explore data in DataFrames and GraphFrames; and 2) we will reveal Spark knowledge graph as a bridge between logical thinking and graph thinking for data mining.

Keywords: Knowledge graph · Spark · Scala · DataFrames · GraphFrames

1 Introduction

On his keynote presentation on Semantics 2017 conference in Amsterdam, Aaron Bradley declared that "Semantic Web has died but in 2012 it was reincarnated by Google Knowledge Graph" [1]. Since that time knowledge graph was adapted by many companies as a powerful way to integrate and search various data. In this paper we will demonstrate some examples showing that knowledge graph abilities are much wider than search and data integration.

Just because Semantic Web was reborn to knowledge graph, does not mean that knowledge graph methodology is limited to Semantic Web methodology. In particularly, knowledge graphs with methods based on SPARQL language have many limitations [2] such as not supported language negation or predicate propertied. In this paper we will show how to build a knowledge graph in Spark Scala and how to explore data using Spark DataFrames and Spark GraphFrames.

Why Spark? Spark is a powerful open source analytics engine with libraries for SQL (DataFrames), machine learning, graphs (GraphFrames) that can be used together for interactive data exploration and supports wide array of data formats and storage systems [3, 4]. Until recently there were no processing frameworks that were able to solve several very different analytical problems like ETL, statistics and graphs in one place. Databricks supports running Spark code on Amazon Cloud via free Databricks Community [5].

DataFrames is a distributed collection of data organized into named columns [6] conceptually the same as table in a relational database. It can be constructed from a

© Springer Nature Switzerland AG 2020
G. Kotsis et al. (Eds.): DEXA 2020 Workshops, CCIS 1285, pp. 96–102, 2020.
https://doi.org/10.1007/978-3-030-59028-4_9

wide variety of sources and was created to leverage power of distributed processing frameworks to solve very complex problems.

Spark GraphFrames [7] is graph library build on top of DataFrames. It is a very powerful tool for performing distributed graph computations on big data that allows to combine graph in relational logic.

2   Artist Biography Knowledge Graph

2.1   Building Knowledge Graph in Spark

After knowledge graph was introduced by Google in 2012 it was adapted by many companies as Enterprise Knowledge Graph and is well known as a powerful engine for search and integration of various data [8].

Knowledge graph concept is much wider than search and data integration. We will show how knowledge graph can be used for data exploration to get a deeper view of data.

To illustrate how knowledge graphs can be built and explored in Spark, as a data domain we will use modern art movement data. First, we will create Artist Biography knowledge graph from a Kaggle dataset 'Museum of Modern Art Collection' [9]. Second, we will find connections between artists based on data from MoMA exhibition: 'Inventing Abstraction 1910–1925' [10]. And third, we will get names of key artists from data about timeline of the 'Modern Art Movements' [11].

From Kaggle dataset we extract biography data (Artist, ArtistBio, Nationality, BeginDate, EndDate, Gender) for several artists:

```
Claude Monet,"(French, 1840-1926)",(French),(1840),(1926),(Male)
Egon Schiele,"(Austrian, 1890-1918)",(Austrian),(1890),(1918),(Male)
Franz Marc,"(German, 1880-1916)",(German),(1880),(1916),(Male)
Georges Braque,"(French, 1882-1963)",(French),(1882),(1963),(Male)
Henri Matisse,"(French, 1869-1954)",(French),(1869),(1954),(Male)
Jackson Pollock,"(American, 1912-1956)",(American),(1912),(1956),(Male)
Joan Miró,"(Spanish, 1893-1983)",(Spanish),(1893),(1983),(Male)
Kazimir Malevich,"(Russian, born Ukraine. 1878-1935)",(Russian),(1878),
(1935),(Male)
Marc Chagall,"(French, born Belarus. 1887-1985)",(French),(1887),
(1985),(Male)
Max Beckmann,"(German, 1884-1950)",(German),(1884),(1950),(Male)
Natalia Goncharova,"(Russian, 1881-1962)",(Russian),(1881),(1962),
(Female)
Oskar Kokoschka,"(Austrian, 1886-1980)",(Austrian),(1886),(1980),(Male)
Pablo Picasso,"(Spanish, 1881-1973)",(Spanish),(1881),(1973),(Male)
Paul Cézanne,"(French, 1839-1906)",(French),(1839),(1906),(Male)
Paul Gauguin,"(French, 1848-1903)",(French),(1848),(1903),(Male)
Paul Klee,"(German, born Switzerland. 1879-1940)",(German),(1879),
(1940),(Male)
Paul Signac,"(French, 1863-1935)",(French),(1863),(1935),(Male)
Piet Mondrian,"(Dutch, 1872-1944)",(Dutch),(1872),(1944),(Male)
Vasily Kandinsky,"(French, born Russia. 1866-1944)",(French),(1866),
(1944),(Male)
Vincent van Gogh,"(Dutch, 1853-1890)",(Dutch),(1853),(1890),(Male)
```

To build Spark knowledge graph on Artist - Nationality relationships, we will create nodes and edges by loosely following the RDF standards, i.e. (subject, object, predicate) form:

```
val graphNodesNationality=aboutArtist.select("Artist").
  union(aboutArtist.select("Nationality")).distinct.toDF("id")
val graphEdgesNationality=aboutArtist.select("Artist","Nationality").
  toDF("src","dst").withColumn("edgeId",lit("Nationality")).distinct
val graphNationality =
  GraphFrame(graphNodesNationality, graphEdgesNationality)
```

2.2 Knowledge Enrichment

Based on data, we can see that some artists changed their nationalities, for example, Marc Chagall, French, was born in Belarus in 1887. This is confusing because Belarus country was founded in 1990. Usually art museums display countries that artists' places of birth currently belong. To enrich knowledge graph code for artist's country of birth and born nationality, we change current country names to historical country names: changed "Russia", "Ukraine", and "Belarus" to "Russian Empire". Then we map artist's country of birth to artist born nationality (bornNationality). Comparing artist's Nationality with bornNationality gives us a list of artists who changed their nationalities:

```
Artist, bornNationality, Nationality
Paul Klee, Swiss, German
Vasily Kandinsky, Russian, French
Marc Chagall, Russian, French
```

Spark code details can be found in our post [12].

2.3 Transform Tabular Data to Knowledge Graph

Another way to compare Spark knowledge graph method with SPARQL is building a knowledge graph from tabular data. In Spark it is very easy to automate this process. First, we will get column names:

```
val columnList=artistBio.columns
columnList: Array[String] = Array(Artist, BeginDate, EndDate,
Nationality, Gender, bornInCountry, bornNationality)
```

Next, for all columns we will create pair edges {'Artist', 'other columns'} and nodes {'Artist'}, {'other columns'}:

```
var graphEdges: DataFrame = Seq(("","","")).toDF("src","dst","edgeId")
var idx=0
for (column <- columnList) {
  graphNodes=graphNodes.union(artistBio.select(column))
  if (idx>0) {
    graphEdges=graphEdges.union(artistBio.
    select(artistBio.columns(0),column).
    toDF("src","dst").withColumn("edgeId",lit(column)))
  }
  idx=idx+1
}
```

Finally we will build a knowledge graph:

```
val graphNodesArtistBio=graphNodes.filter('id=!="").distinct
val graphEdgesArtistBio=graphEdges.filter('src=!="").distinct
val graphArtistBio =
  GraphFrame(graphNodesArtistBio, graphEdgesArtistBio)
```

Spark code details can be found in our post [12].

2.4 Graph Visualization

To visualize graph (Fig. 1) we will translate graph edges to DOT language - graph description language for graph visualization [13]. For graph visualization we used Gephi tool [14].

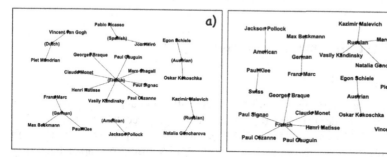

Fig. 1. Subgraphs on Artist Bio knowledge graph: a) Artists Nationalities from Kaggle dataset; b) Artists Born Nationalities.

Graph edges to DOT language code:

```
display(graphArtistBio.edges.filter('edgeId==="bornNationality").
  map(s=>("\""+s(0).toString +"\" -> \""+s(1).toString +"\""+" ;")))
"Natalia Goncharova" -> "Russian" ;
"Georges Braque" -> "French" ;
"Egon Schiele" -> "Austrian" ;
"Franz Marc" -> "German" ;
"Paul Klee" -> "Swiss" ;
"Marc Chagall" -> "Russian" ;
"Joan Miró" -> "Spanish" ;
"Vincent van Gogh" -> "Dutch" ;
"Kazimir Malevich" -> "Russian" ;
"Pablo Picasso" -> "Spanish" ;
"Oskar Kokoschka" -> "Austrian" ;
"Vasily Kandinsky" -> "Russian" ;
```

2.5 Knowledge Graph Queries

Here are two examples of knowledge graph queries that are using Spark instead of 'traditional' SPARQL language. Let's say we need to find pairs of artists that were born in the Austria.

First method is using Spark DataFrames language by self-joining ArtistBio table:

```
val sameCountryBirthDF = artistBio.select("Artist","bornInCountry").
  join(artistBio.select("Artist","bornInCountry").
  toDF("Artist2","bornInCountry2"),'bornInCountry==='bornInCountry2).
  filter('Artist<'Artist2).
  select("bornInCountry","Artist","Artist2").distinct
Austria, Egon Schiele, Oskar Kokoschka
```

Second method is using Spark GraphFrames motif 'find' function [15]. This method is conceptually similar to {subject - predicate -> object} and it is better understandable and more elegant than self-joining of tabular data:

```
val sameCountryBirthGF=graphArtist.
  find("(a) - [ac] -> (c); (b) - [bc] -> (c)").
  filter($"ac.edgeId"===$"bc.edgeId" && $"ac.edgeId"==="bornInCountry").
  filter($"a.id"<$"b.id").select("c.id","a.id","b.id").
  toDF("bornInCountry","Artist1","Artist2").distinct
```

Austria, Egon Schiele, Oskar Kokoschka

3 Modern Art Movement Knowledge Graph

3.1 Semi-structured Data

To show how to use knowledge graph to integrate different data, first we will build a knowledge graph of modern art key artists. From semi-structured dataset about timeline of the Modern Art Movements [11] we will get names of key artists of modern art movements. From this list we will get a subset of artists from our artist biography knowledge graph. In our post [16] you can find data processing code in details.

```
1872 - 1892, Impressionism, Claude Monet
Early 1880s - 1914, Post-Impressionism, Paul Gauguin
Early 1880s - 1914, Post-Impressionism, Paul Signac
Early 1880s - 1914, Post-Impressionism, Paul Cézanne
1905 - 1910, Fauvism, Henri Matisse
1907 - 1922, Cubism, Pablo Picasso
1907 - 1922, Cubism, Georges Braque
1909 - late 1920s, Futurism, Natalia Goncharova
1913 - late 1920s, Suprematism, Kazimir Malevich
1917 - 1931, De Stijl, Piet Mondrian
1924 - 1966, Surrealism, Joan Miró
1943 - 1965, Abstract Expressionism, Jackson Pollock
```

Knowledge graph edges:

```
val modernArtEdges=
  modernArtData.select("keyArtist","artMovement").toDF("src","dst").
  withColumn("edgeId",lit("artMovement")).
  union(modernArtData.select("artMovement","time").toDF("src","dst").
  withColumn("edgeId",lit("time"))).distinct
```

Modern Art Movement knowledge graph:

```
val modernArtGraph=GraphFrame(modernArtEdges.select("src").
  union(modernArtEdges.select("dst")).distinct.toDF("id"),
  modernArtEdges.distinct)
```

3.2 Knowledge Graph Integration

There are two ways to integrate two knowledge graphs: combine edges of both graphs or overlap vertices of both graphs. To integrate Modern Art Movement knowledge graph with Artist Biography knowledge graph we will follow the second way.

From Artist Biography knowledge graph we will take only information about nationalities and countries where artists were born.

```
val modernArtBioEdges = graphArtistBio.edges.
  join(modernArtKeyArtists,'src==='keyArtist).drop("keyArtist").
  union(modernArtGraph.edges).
  filter(not('edgeId.isin("EndDate","BeginDate","Gender")))
```

Build a knowledge graph:

```
val modernArtBioGraph=GraphFrame(
  modernArtBioEdges.select("src").
  union(modernArtBioEdges.select("dst")).distinct.toDF("id"),
  modernArtBioEdges)
```

Integrated knowledge graph shows biographies of Modern Art Movements key artists (Fig. 2).

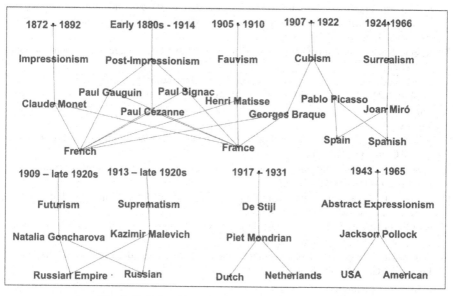

Fig. 2. Relationships between modern art movements.

This graph shows unknown connections between modern art movements: Impressionism, Post-Impressionism, Fauvism, Cubism and Surrealism were created by artists born in France or Spain. These art movements had no connections with Futurism and Suprematism that were created by artists born in Russian Empire.

4   Conclusion

In this paper we illustrated how Spark knowledge graph can build a bridge between logical thinking and graph thinking for data exploration. In data mining examples we demonstrated that Spark GraphFrames library provides a flexibility to switch between SQL functions and graph functions.

Exploring data about modern art artists we found unknown connections between artists and between modern art movements.

References

1. Bradley, A.: Semantics 2017. https://2017.semantics.cc/aaron-bradley-eamonn-glass
2. The Limitations of SPARQL. http://horicky.blogspot.com/2010/08/limitations-of-sparql.html
3. Apache Spark. https://databricks.com/spark/about
4. Chambers, B., Zaharia, M.: Spark: The Definitive Guide: Big Data Processing Made Simple
5. Databricks Community Edition. https://databricks.com/blog/2016/02/17/introducing-databricks-community-edition-apache-spark-for-all.html
6. Spark DataFrames. https://databricks.com/blog/2015/02/17/introducing-dataframes-in-spark-for-large-scale-data-science.html
7. Spark GraphFrames. https://databricks.com/blog/2016/03/03/introducing-graphframes.html
8. Industry-scale Knowledge Graphs: Lessons and Challenges. https://queue.acm.org/detail.cfm?id=3332266
9. Kaggle dataset 'Museum of Modern Art Collection'. https://www.kaggle.com/momanyc/museum-collection
10. MoMA exhibition: Inventing Abstraction 1910–1925. https://www.moma.org/interactives/exhibitions/2012/inventingabstraction/?page=artists
11. Timeline of the 'Modern Art Movements'. https://drawpaintacademy.com/modern-art-movements/
12. "Knowledge Graph for Data Mining" post. http://sparklingdataocean.com/2019/09/24/knowledgeGraphDataAnalysis/
13. Drawing graphs with dot. https://www.ocf.berkeley.edu/~eek/index.html/tiny_examples/thinktank/src/gv1.7c/doc/dotguide.pdf
14. Visual network analysis with Gephi. https://medium.com/@EthnographicMachines/visual-network-analysis-with-gephi-d6241127a336
15. Motifs Findings in GraphFrames. https://www.waitingforcode.com/apache-spark-graphframes/motifs-finding-graphframes/read
16. "Knowledge Graph for Data Integration" post. http://sparklingdataocean.com/2020/02/02/knowledgeGraphIntegration/

Open Information Extraction
for Knowledge Graph Construction

Iqra Muhammad[(✉)], Anna Kearney, Carrol Gamble, Frans Coenen,
and Paula Williamson

Department of Computer Science, The University of Liverpool,
Liverpool L693BX, UK
{iqra,frans.coenen}@liverpool.ac.uk

Abstract. An open information extraction approach for knowledge
graph construction is presented. The motivation for the work is that large
quantities of scholarly documents are available within many domains of
discourse, and the subsequent challenge is to identify the most relevant
articles concerning a particular topic. The proposed approach takes a
document corpus and identifies triples within this corpus which are then
processed to generate a literature knowledge graph. The extraction of
triples is conducted using an open information extraction approach. The
proposed OIE4KGC approach was evaluated using a bespoke clinical
research methodology dataset and a benchmark dataset. A f-score of 51%
was achieved on a clinical research methodology dataset and a f-score of
37% was achieved on the benchmark dataset.

Keywords: Open information extraction · Literature knowledge graph
construction

1 Introduction

The number of available scientific papers has been increasing at an exponential
rate. In 2009, it was estimated that the 50 million mark in the number of scientific papers was passed [2]. One solution is online article repositories, which typically feature some limited form of search facility. One example is the abstracts
stored in the MEDLINE[1] repository, which can be accessed (searched) using
the PubMed[2] interface. However the search functionality supported by these
systems is typically inadequate for efficiently searching large repositories. An
alternative solution, which is advocated in this paper, is to store the document
corpus in a literature knowledge graph [14,15] where the vertices represent concepts and documents, and the edges represent relationships between concepts,
or concepts and documents [1,9–11,18,19]. This, it is suggested, will provide
a better organisation of the data and consequently provide for more effective
information retrieval and knowledge understanding.

---

[1] https://www.nlm.nih.gov/bsd/medline.html.
[2] https://www.ncbi.nlm.nih.gov/pubmed/.

© Springer Nature Switzerland AG 2020
G. Kotsis et al. (Eds.): DEXA 2020 Workshops, CCIS 1285, pp. 103–113, 2020.
https://doi.org/10.1007/978-3-030-59028-4_10

To create a literature knowledge graph, information extraction techniques are applied to the unstructured text in the document corpus. The extracted information can then be used to build the desired literature knowledge graph. However, there are many challenges to building effective information retrieval systems regardless of the domain of discourse. A particular issue is that the vocabulary in many domains tends to be extensive, compounded by the fact that there are often semantic variations for the same concept and that the relationships between concepts are often complex. This is especially the case in the clinical domain.

Traditional information extraction techniques for building knowledge graphs tend to use a pre-defined schema, an agreed set of specific concept types and relation types for vertices and relations; and typically operate using domain-specific supervised learning approaches that require training data [12, 22–25]. However, the training data and a schema specific to a domain of discourse is typically not available in many cases. An alternative is to use domain independent Open Information Extraction (OIE) models that are already pre-trained on general datasets. The knowledge graphs constructed using OIE do not require a pre-defined schema. Open information extraction techniques make use of a set of patterns to extract triples. Each triple consists of two arguments, a subject and an object, and a predicate (relation) linking the arguments, which can then be used to construct a knowledge graph [8].

This paper presents the Open Information Extraction for Knowledge Graph Construction (OIE4KGC) approach; a novel process for generating a literature knowledge graph from a given corpus using the concept of OIE. The focus for the work is clinical trial's methodology literature; an essential resource in facilitating clinical trials research. The dataset used for evaluation purposes consisted of 400 abstracts on recruitment strategies for clinical trials, selected from the ORRCA dataset[3] [4]. The abstracts in this dataset represent recruitment strategies, adopted by clinical trials, when recruiting patients for trials.

The rest of this paper is structured as follows. Section 2 considers previous work directed at the concept of creating knowledge graphs from unstructured text. In Sect. 3 the proposed OIE4KGC approach is described. Section 4 then presents the evaluation of the proposed OIE technique for knowledge graph generation. Finally, Sect. 5 concludes the paper with a summary of the main findings and directions for future work.

2   Literature Review

This literature review section presents an overview of existing work on open information extraction and knowledge graph construction relevant to the work presented here. It has been divided into two sections. The first, Sect. 2.1 considers

---

[3] The ORRCA (Online Resource for Recruitment Research in Clinical Trials) dataset is part of a PhD with the University of Liverpool's Biostatistic's department. This dataset will be released publicly on the author's website.

OIE techniques; and the second, Sect. 2.2, describes some of the existing work on knowledge graphs.

2.1 Open Information Extraction

Recently, many attempts have been made at using OIE techniques for converting unstructured text to structured text. Existing techniques based on the idea of OIE use a set of patterns to convert a sentence into relational triples. OIE techniques can be divided into three categories: (i) learning-based, (ii) rule-based and (iii) inter-proposition-based:

Learning-Based Systems. Learning-based open information extraction systems use training data, from which a model is learned for producing relational triples. One of the first systems directed at learning-based OIE was TextRunner [5]. Using TextRunner, a small sample of sentences are first parsed using Penn Treebank after which a dependency parser is applied to identify and label a set of "extractions" as positive and negative training examples. In [27] an open information extraction system is used, and relies on a bootstrapping approach based on a wikipedia dataset [6]. In [8] an OIE system, called RnnOIE[4] founded on a deep-learning based approach was presented. RnnOIE is a model pre-trained on the OIE2016 dataset. The reported experiments demonstrated that RnnOIE, was able to outperform many state of the art benchmarks. The RnnOIE tool was therefore adopted with respect to the OIE4KGC approach described in this paper.

Rule-Based Systems. A number of approaches to OIE make use of hand-crafted extraction rules. One example is REVERB [7], this is a "shallow extractor" that makes use of hand-craft extraction rules. REVERB addresses the problems of uninformative and incoherent extractions. Another rule-based approach is PredPratt [16], which used a set of non-lexicalised rules, defined over universal dependency parses, for extracting predicate-argument structures. The disadvantage is the need to hand-craft the rule set. The approach was therefore deemed inappropriate for the knowledge graph generation application.

Inter-Proposition Relationship Based Systems. OIE systems extract a list of relational triples also called propositions, where each proposition consists of a single predicate and a number of arguments from an input sentence. The majority of the above mentioned OIE systems are not capable of capturing the complete expression in a sentence as they ignore the context under which a proposition is complete. An example of such a scenario is the relational triple ⟨*Barack Obama, was a, good president*⟩ from the sentence "Democrats believe that Barack Obama was a good president"; this triple is inappropriate since the input sentence is not asserting this proposition. Such shortcomings can be handled by OLLIE [26], which adds an additional attribute context to the extracted relation triple or proposition, showing that a proposition is reported by some entity (*AttributedTo believe; Democrats*). This idea of adding additional attributes to

---

[4] https://github.com/gabrielStanovsky/supervised-oie.

an extracted relation triple or proposition is referred to as inter-propositional relation. Another similar state-of-the-art approach was proposed in [13] where a nested representation for OIE was presented. This approach was able to capture high-level dependencies, allowing for an improved representation of the meaning of a sentence. However, for the purpose of building literature knowledge graphs this limitation of learning based systems was accepted; additional attributes could always be added at a later date.

2.2 Knowledge Graphs

As noted in the introduction to this paper, knowledge graphs are labelled graphs where vertices represent concepts and edges represent relations between them. Previous work on the automatic construction of knowledge graphs can be divided into two categories: (i) Domain-Specific Knowledge graphs and (ii) Literature Knowledge graphs:

Domain Specific Knowledge Graphs. Domain Knowledge Graphs, as the name implies, are domain-specific, meaning that the text used in the construction of the knowledge graph is limited to a specialised field like biology, physics, computer science or any other domain of discourse. One of the first few attempts at creating a knowledge graph in the biomedical science domain involved the use of rdf-extraction from excel sheets in [20]. A recent, frequently cited work [21] focused on the construction of a knowledge graph for the domain of biomedical sciences.

Literature Knowledge Graphs. Literature knowledge graphs act as a storage mechanism for representing concepts and relations in the literature associated with some domain of interest. A well-known literature knowledge graph, is that used within Semantic Scholar is presented in [15]. Another well-known literature knowledge graph was created by Microsoft and comprised author vertices, concept vertices, paper vertices and edges connecting them [28].

3 Open Information Extraction for Knowledge Graph Construction (OIE4KGC)

In this section the proposed OIE4KGC approach is presented. Subsect. 3.1 first gives a problem definition for literature knowledge graph construction. There are two kinds of vertices in the envisioned knowledge graph: (i) concept vertices and (ii) document vertices. The vertices are linked by edges representing relationships. The proposed approach is illustrated in Fig. 1 with an example taken from the ORRCA dataset used for evaluation purposes. From the figure it can be seen that OIE4KGC comprises four main components: (i) Triple Extraction. (ii) Triple Filtering, (iii) Concept Linking and (iv) Merging of vertices and Knowledge graph population. Each is discussed in further detail in Subsects. 3.2, 3.3 and 3.4. The pseudocode for the OIE4KGC is given in Algorithm 1; this will be referred to in the following sub-sections.

3.1 Problem Definition

The aim is to construct a literature knowledge graph $G = \{V, E\}$ where the set of vertices V represent documents (abstracts) or concepts, and the set of edges E represent relationships. Given, a corpus of n documents (abstracts), $D = \{D_1, \ldots, D_n\}$, where each document is comprised of m sentences $S = \{S_1, \ldots, S_m\}$, the task is to find an appropriate set of triples T from each sentence in each document and use these triples to construct G. A triple T takes the form $\langle a_s, r, a_o \rangle$, where a_s is the subject argument, a_o is the object argument and r is a predicate (relation) between them; each is represented by a string. The arguments represent concepts which may potentially be included in the eventual knowledge graph.

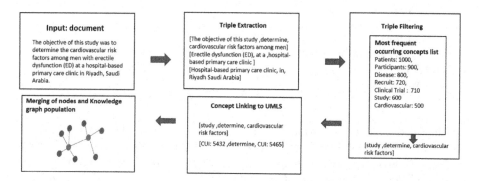

Fig. 1. Stages involved in the construction of a literature knowledge graph using OIE4KGC

3.2 Triple Extraction

Triple Extraction is the first stage in the proposed approach. A variety of tools are available that can be used to extract triples from unstructured text, both supervised and unsupervised. The assumption was, as in the case of the clinical research methodology scenario used as the focus for this paper, that training data was not available. Hence a semi-supervised approach was required; a pre-trained OIE tool of the form discussed in Sect. 2. For the proposed approach RnnOIE [8] was used because as reported in [8], RnnOIE had been shown to outperform other state of the art tools for OIE using benchmark datasets.

The first step in the application of any OIE tool is the pre-processing of the data so as to identify sentences, $S = \{S_1, \ldots, S_m\}$ (line 6 in the pseudo code given in Algorithm 1) for every document D_i in the corpus D (line 4 in Algorithm 1). With respect to the proposed approach the Spacy's sentence segmentation tool was used[5]. The second step (line 9) was to apply RnnOIE to each sentence [8]. In this manner the predicates and arguments in each sentence could

---

[5] https://spacy.io/.

Algorithm 1. OIE for Knowledge graph Pseudocode

1: Input D, Output G
2: D = A set of Documents, G = Empty Knowledge graph database
3: L = Lexicon of most frequently occurring words in D
4: **for** $D = \{1, 2, \ldots, i\}$ **do**
5:     $G = G$ plus vertex representing D_i
6:     $S =$ Set of Sentences in D_i
7:     $T = \oslash$ (Set to hold triples)
8:     **for** $S = \{1, 2, \ldots, i\}$ **do**
9:         $T_i =$ Set of triples in S_i
10:        **for** $t = \{1, 2, \ldots, i\}$ **do** where $t_i = \langle a_s, r, a_o \rangle$ in T_i
11:            $t_i = t_i$ with noun chunking applied
12:            $t_i = t_i$ with only nouns that are retained after checking L
13:            $t_i = t_i$ annotated with CUI
14:        **end for**
15:        $T = T \cup T_i$
16:    **end for**
17:    $G = G$ incorporating $T = \{T_1, \ldots, T_i\}$
18: **end for**
19: Exit with G

---

be identified without requiring any domain specific knowledge. In Fig. 1 three triples are identified. The first of these ⟨*the objective of this study, determine, cardiovascular risk factors among men*⟩, where *determine* is the relation (predicate), and *the objective of this study* and *cardiovascular risk factors among men* are its arguments. The identified argument and relation strings, as illustrated in the example, were expected to include unnecessary words which should be removed. This was done (line 11) using Spacy's Noun Chunker so that only the informative noun phrases for arguments were retained. Thus, the above example will be reduced to ⟨*study, determine, cardiovascular risk factors*⟩.

3.3   Triple Filtering

The aim of the triple filtering stage was to filter the triples extracted from each of the abstracts in a given corpus during the Triple Extraction Stage (Stage 1) and removed redundant and uninformative words within arguments. As a result some arguments would be "empty". Informative words were considered to be words that appear frequently in the given corpus. To this end a "Most frequent occurring concepts list" was constructed; a lexicon L of the 100 most frequently occurring nouns in the corpus. A fragment of such a lexicon is given in Fig. 1; alongside each entry is its associated occurrence count. For each argument in each triple the words appearing in L were retained (line 12). In Fig. 1 the triple ⟨*study, determine, cardiovascular risk factors*⟩ remains unchanged because all these word are present in L. The objective of the triple filtering stage was also to ensure, that if the triples are converted to knowledge graph embeddings, the embeddings would not be sparse.

3.4   Linking of Clinical Concepts to UMLS

The arguments contained in the triples retained after Stage 2 express concepts to be included in the knowledge graph. Several arguments might express the same concept, whilst at the same time a single argument can express different concepts depending on context. To construct a useful knowledge graph these ambiguities need to be resolved. The central idea for achieving this was to annotate each argument with additional information, for example with its synonyms and/or hyponyms, that would allow disambiguation. In the case of the clinical research methodologies application domain this was achieved by annotating each argument with the relevant Concept Unique Identifier (CUI) held in the Unified Medical Language System (UMLS) Metathesaurus [3]. Using the words and phrases held in the metathesaurus, the arguments in the identified triples were annotated with a CUI indicating the sense of the argument (line 13). For example the word "study" has the CUI 5432, while the phrase "cardiovascular risk factors" has the CUI 5465 (as indicated in Fig. 1). These CUI annotations were then used for disambiguating purposes in Stage 4.

3.5   Merging of Vertices and Knowledge Graph Population

The final stage in the proposed approach is the construction of the desired knowledge graph (line 17 in the pseudo code). A knowledge graph can be represented using a variety of graph database models, with respect to the work presented in this paper Neo4j was used[6]. Custom data-structures were created for concept vertices, document vertices and for relations. The arguments within each triple represent concepts to be included in the knowledge graph. Figure 2 shows a toy example of a literature knowledge graph generated using the proposed OIE4KGC approach. In the figure there are two types of vertices: (i) concept vertices (blue) and (ii) document vertices (green). Each concept vertex has two properties, (i) the argument string (the concept name) and (ii) the associated CUI based ID that links the argument string to the UMLS Metathesaurus sense (included to add additional information). Each document vertex references a document (abstract). A document vertex in the knowledge graph also has two properties: (i) the title string of the abstract and (ii) a unique identifier (it cannot be assumed that each document will have a unique title). There are two kinds of edge in the knowledge graph: (i) edges linking concepts and (ii) edges linking concepts and documents. Edges linking documents and concepts have the label "mentions" indicating that the document mentions the indicated concept. Edges linking a pair of concepts indicate relations extracted using OIE.

To generate the literature knowledge graph each triple was processed in turn. For each triple two new vertices were created, v_s and v_o, connected by the given relation r, and each connected to the document vertex created for D_i. These were then compared to the knowledge graph G so far. There are four options:

1. If v_s and v_o match two vertices v_1 and v_2 in G: merge v_s and v_o with v_1 and v_2 adding the relation r if not already in existence.

---

[6] https://neo4j.com/.

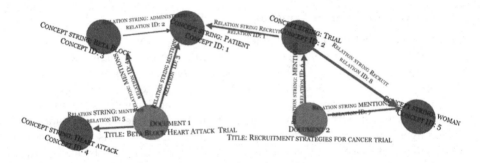

Fig. 2. A toy example of a literature knowledge graph generated using OIE4KGC (Color figure online)

2. If v_s matches a vertex v_1 in G, but v_o does not match any vertex in G: merge v_s with v_1.
3. If v_o matches a vertex v_2 in G, but v_s does not match any vertex in G: merge v_o with v_2.
4. Otherwise (v_s and v_o do not match any vertices in G): do nothing.

To facilitate the merging Neo4j has a merge utility.

4   Evaluation

This section describes the evaluation conducted to assess the performance of the proposed approach. The evaluation was centred on the RnnOIE OIE tool [8] central to the proposed OIE4KGC approach. For the evaluation two data sets were used: (i) the ORRCA data set [4] and (ii) the Reverb ClauseIE dataset[7] [17]. For the ORRCA data set 100 sentences were randomly chosen and a "gold standard2" set of triples identified by manual inspection of the 100 sentences. The ClauseIE data set is a benchmark dataset of 500 sentences manually labelled for OIE; 100 sentences were randomly selected from the ClauseIE data set. The evaluation metric used was F-score, the harmonic mean of precision and recall. Note that precision was defined as the number of correct triples divided by the total number of triples extracted by RnnIE tool, whilst recall was defined as the number of correct triples divided by the number of triples in the gold standard for selected 100 sentences. It should be noted that the objective of this evaluation was to assess the RnnOIE tool at the sentence level.

The results obtained are given in Table 1. From the table it can be seen RnnOIE was able to achieve an F-score of 51% using the ORRCA data set and 37% that using the ClauseIE dataset. It can also be seen from Table 1 that the precision was better using the ORRCA dataset compared to the ClauseIE

---

[7] https://www.mpi-inf.mpg.de/departments/databases-and-information-systems/soft-ware/clausie/.

dataset. This difference in precision can be accounted for by the structural differences in sentences in both datasets. Triples extracted from the ClauseIE dataset have numerical values in the arguments; which, using the proposed approach, results in a triple being discarded. It is also note-worthy that the sentences in the ORRCA dataset are longer than in the case of the ClauseIE dataset; the average number of words in each sentence for the ORRCA dataset was 30 compared to 10 for the ClauseIE dataset. From the results it can be concluded that RnnOIE is appropriate for clinical document collections as exemplified by the ORRCA dataset, and appropriate for inclusion in the proposed OIE4KGC approach advocated in this paper.

Table 1. Table showing the performance of the RnnIE tool on the ORRCA and ClauseIE datasets

Dataset	Precision	Recall	F-score
ClauseIE dataset	0.473	0.311	0.375
ORRCA dataset	0.783	0.391	0.512

5   Conclusion and Future Work

This paper has presented the Open Information Extraction for Knowledge Graph Construction (OIE4KG) approach for constructing literature knowledge graphs. The focus of the work was a clinical trials methodological articles collection. Open information extraction was used for the extraction of triples from the document collection. The RnnOIE too was evaluated using two datasets, ORRCA and ClauseIE. The F-score of 51% percent using the ORRCA dataset suggests that OIE tools such as RnnOIE can be successfully used to construct literature knowledge graphs in the clinical domain. In terms of future research, the intention is to focus on canonicalizing the knowledge graph and using the knowledge graph embeddings for tasks like document retrieval and document ranking.

References

1. Luan, Y., He, L., Ostendorf, M., Hajishirzi, H.: Multi-task identification of entities, relations, and coreference for scientific knowledge graph construction (2018)
2. Jinha, A.E.: Article 50 million: an estimate of the number of scholarly articles in existence. Learn. Publ. **23**, 258–263 (2010)
3. Bodenreider, O.: The Unified Medical Language System (UMLS): integrating biomedical terminology. Nucl. Acids Res. **32**, D267–D270 (2004)
4. Kearney, A., et al.: Development of an online resource for recruitment research in clinical trials to organise and map current literature. Clin. Trials **15**, 533–542 (2018)

5. Yates, A., Cafarella, M., Banko, M., Etzioni, O., Broadhead, M., Soderland, S.: TextRunner. In: Proceedings of Human Language Technologies: The Annual Conference of the North American Chapter of the Association for Computational Linguistics: Demonstrations on XX - NAACL 2007 (2007)
6. Weld, D.S., Hoffmann, R., Wu, F.: Using Wikipedia to bootstrap open information extraction. ACM SIGMOD Rec. **37**, 62 (2009)
7. Fader, A., Zettlemoyer, L.: Paraphrase-driven learning for open question answering. In: Proceedings of the 51st Annual Meeting of the Association for Computational Linguistics (Volume 1: Long Papers), pp. 1608–1618 (2013)
8. Stanovsky, G., Michael, J., Zettlemoyer, L., Dagan, I.: Supervised open information extraction. In: Proceedings of the 2018 Conference of the North American Chapter of the Association for Computational Linguistics: Human Language Technologies, Volume 1 (Long Papers) (2018)
9. Cui, L., Wei, F., Zhou, M.: Neural open information extraction. In: Proceedings of the 56th Annual Meeting of the Association for Computational Linguistics (Volume 2: Short Papers) (2018)
10. Zhan, J., Zhao, H.: Span model for open information extraction on accurate corpus. https://arxiv.org/abs/1901.10879
11. Jiang, M., et al.: MetaPAD. In: Proceedings of the 23rd ACM SIGKDD International Conference on Knowledge Discovery and Data Mining - KDD 2017 (2017)
12. Qin, L., Hao, Z., Yang, L.: Question answering system based on food spot-check knowledge graph. In: Proceedings of 2020 the 6th International Conference on Computing and Data Engineering (2020)
13. Bhutani, N., Jagadish, H.V., Radev, D.: Nested propositions in open information extraction. In: Proceedings of the 2016 Conference on Empirical Methods in Natural Language Processing (2016)
14. Jaradeh, M.Y., et al.: Open research knowledge graph. In: Proceedings of the 10th International Conference on Knowledge Capture - K-CAP 2019 (2019)
15. Ammar, W., et al.: Construction of the literature graph in semantic scholar. In: Proceedings of the 2018 Conference of the North American Chapter of the Association for Computational Linguistics: Human Language Technologies, Volume 3 (Industry Papers) (2018)
16. White, A.S., et al.: Universal decompositional semantics on universal dependencies. In: Proceedings of the 2016 Conference on Empirical Methods in Natural Language Processing (2016)
17. Corro, L.D., Gemulla, R.: ClausIE. In: Proceedings of the 22nd International Conference on World Wide Web - WWW 2013 (2013)
18. Zhao, S., Su, C., Sboner, A., Wang, F.: Graphene. In: Proceedings of the 28th ACM International Conference on Information and Knowledge Management - CIKM 2019 (2019)
19. Huang, Z., Yang, J., van Harmelen, F., Hu, Q.: Constructing knowledge graphs of depression. In: Siuly, S., Huang, Z., Aickelin, U., Zhou, R., Wang, H., Zhang, Y., Klimenko, S. (eds.) HIS 2017. LNCS, vol. 10594, pp. 149–161. Springer, Cham (2017). https://doi.org/10.1007/978-3-319-69182-4_16
20. Han, L., Finin, T., Parr, C., Sachs, J., Joshi, A.: RDF123: from spreadsheets to RDF. In: Sheth, A., et al. (eds.) ISWC 2008. LNCS, vol. 5318, pp. 451–466. Springer, Heidelberg (2008). https://doi.org/10.1007/978-3-540-88564-1_29
21. Belleau, F., Nolin, M.-A., Tourigny, N., Rigault, P., Morissette, J.: Bio2RDF: towards a mashup to build bioinformatics knowledge systems. J. Biomed. Inform. **41**, 706–716 (2008)

22. Haussmann, S., et al.: FoodKG: a semantics-driven knowledge graph for food recommendation. In: Ghidini, C., et al. (eds.) ISWC 2019. LNCS, vol. 11779, pp. 146–162. Springer, Cham (2019). https://doi.org/10.1007/978-3-030-30796-7_10
23. Luan, Y., He, L., Ostendorf, M., Hajishirzi, H.: Multi-task identification of entities, relations, and coreference for scientific knowledge graph construction. In: Proceedings of the 2018 Conference on Empirical Methods in Natural Language Processing (2018)
24. Wadden, D., Wennberg, U., Luan, Y., Hajishirzi, H.: Entity, relation, and event extraction with contextualized span representations. In: Proceedings of the 2019 Conference on Empirical Methods in Natural Language Processing and the 9th International Joint Conference on Natural Language Processing (EMNLP-IJCNLP) (2019)
25. Silva, V., Freitas, A., Handschuh, S.: Building a Knowledge Graph from Natural Language Definitions for Interpretable Text Entailment Recognition. https://www.aclweb.org/anthology/L18-1542/
26. Schmitz, M.: Open Language Learning for Information Extraction (2010). https://www.aclweb.org/anthology/D12-1048.pdf
27. Weld, D., Hoffmann, R., Wu, F.: Using Wikipedia to bootstrap open information extraction. ACM SIGMOD Rec. **37**, 62 (2009)
28. Microsoft Academic Knowledge Graph. http://ma-graph.org/

Author Index

Printed in the United States
By Bookmasters